Arthropod Pests of Horticultural Crops in Tropical Asia

Arthropod Pests of Horticultural Crops in Tropical Asia

Rangaswamy Muniappan

Virginia Tech, Blacksburg, Virginia, USA

B. Merle Shepard

Clemson University, Coastal Research and Education Center
Charleston, South Carolina, USA

Gerald R. Carner

Clemson University, Clemson, South Carolina, USA

and

Peter Aun-Chuan Ooi

Universiti Tunku Abdul Rahman (UTAR) Perak Campus,
Kampar, Perak Darul Ridzuan, Malaysia

www.cabi.org

Integrated Pest Management Collaborative Research Support Program IPM CRSP

United States Agency for International Development USAID

USAID FROM THE AMERICAN PEOPLE

CABI is a trading name of CAB International

CABI
Nosworthy Way
Wallingford
Oxfordshire OX10 8DE
UK

Tel: +44 (0)1491 832111
Fax: +44 (0)1491 833508
E-mail: info@cabi.org
Website: www.cabi.org

CABI
875 Massachusetts Avenue
7th Floor
Cambridge, MA 02139
USA

Tel: +1 800 552 3083 (toll free)
Fax: +1 (0)617 395 4051
E-mail: cabi-nao@cabi.org

A catalogue record for this book is available from the British Library, London, UK.

Library of Congress Cataloging-in-Publication Data

Arthropod pests of horticultural crops in tropical Asia / Rangaswamy Muniappan ... [et al.].
p. cm.
Includes bibliographical references and index.
ISBN 978-1-84593-951-9 (alk. paper)
1. Horticultural crops--Diseases and pests--Asia. 2. Arthropod pests--Asia. I. Muniappan, R.

SB608.H83A78 2012
635.9'2095--dc23

2012001018

ISBN-13: 978 1 84593 951 9

Commissioning editor: Rachel Cutts
Editorial assistant: Alexandra Lainsbury
Production editor: Fiona Chippendale

Typeset by Columns Design XML Ltd, Reading
Printed and bound by Gutenberg Press Limited, Tarxien, Malta

Contents

Authors

Rangaswamy Muniappan, *Program Director, IPM CRSP, Virginia Tech, 526 Prices Fork Road, Blacksburg, Virginia 24061, USA.* E-mail: rmuni@vt.edu

B. Merle Shepard, *Emeritus Professor of Entomology, Clemson University, Coastal Research and Education Center, 2700 Savannah Hwy, Charleston, South Carolina 29414, USA.* E-mail: mshprd@clemson.edu

Gerald R. Carner, *Emeritus Professor of Entomology, School of Agricultural Forest and Environmental Sciences, 114 Long Hall, Clemson University, Clemson, South Carolina 29634, USA.* E-mail: gcarner@clemson.edu

Peter Aun-Chuan Ooi, *Faculty of Science, Tan Sri Dato' Philip Kuok Professorial Chair for Agricultural Sciences, Universiti Tunku Abdul Rahman (UTAR) Perak Campus, Jalan Universiti, Bandar Barat, 31900 Kampar, Perak Darul Ridzuan, Malaysia.* E-mail: peteracooi@gmail.com & ooiac@utar.edu.my

Foreword

The world's population surpassed 7 billion in late 2011, with predictions that it will exceed 9 billion by 2050. Demand for food from cereal and horticultural crops is expected to grow accordingly. As many Asian countries become more developed, the demand for food from horticultural crops is expected to increase even more. Agriculture will be required to meet this food demand with less land, less water, and less labor. Although staple food is provided by the two main cereals, rice and wheat, food from horticultural crops such as beans, vegetables, and fruits is also significant for humans. The huge variety of food produced from horticultural crops will provide not only diversity in taste but also sources of nutrients, minerals, and vitamins for humankind to enjoy healthy food. Nonetheless, insect pests are constant threats to horticultural crop production. Insects are the principal causes of yield losses in cabbage, mango, and bean crops.

In Asia, growers generally rely on insecticides to avoid crop loss. In the past 5 years, insecticide usage in many developing countries has escalated and misuse is becoming more abundant. Pesticide residues in horticultural crops are becoming a deep concern. For growers to manage their crops properly with a minimal use of pesticides, it is vitally important to be able to diagnose pest problems. Similarly, for crop protection technicians and educators to be able to advise growers and teach, they need to identify insect pests, assess the damage they cause, and identify the related natural enemies.

This book, *Arthropod Pests of Horticultural Crops in Tropical Asia,* is an excellent reference for researchers, extension personnel, and university students. The text provides detailed descriptions of pests, their taxonomy, Latin and common names, their host ranges, the kinds of damage they cause, and their natural enemies. Most relevant of all is the excellent photography by the authors of the pest species and damage for readers to view. The photographs are from the authors' decades-old photo collections and many patient hours spent in capturing the images. I am very glad that the authors are sharing their valuable photos with the world. This book will be an important contribution to science, research, and the implementation of integrated pest management of horticultural crops in Asia.

Robert S. Zeigler
Director General
International Rice Research Institute

Acknowledgements

The authors are grateful to the USAID-funded project Agreement No. EPP-A-00-04-00016-00 Integrated Pest Management Collaborative Research Support Program (IPM CRSP) for its support in the publication of this book. We thank Miriam Rich, Melissa Smith, and Kellsey Lequick for their efforts in editing the narrative and Mark (Amado) Ohland for editing photos.

1 Arthropod Pests and their Natural Enemies on Horticultural Crops in Tropical Asia

About one-third of the human population and half the world's poor and malnourished live in tropical Asia. Agriculture is the backbone of this region, but it still lags behind other regions, except tropical Africa, in yields per hectare. Research on cereal crops in the 1960s led to the green revolution with the development of high-yielding varieties. With these varieties, food production and self sufficiency did occur in India in the 1970s and 1980s; however, stagnation in production in the 1990s and 2000s rekindled food security concerns in the region. In the green revolution, most of the emphasis was on high-yielding cereal grains such as rice, wheat, and corn, while horticultural crops received little attention for their role in food security and reduction of poverty and malnutrition. Sustainable horticultural crop production should be considered in association with other agricultural crop production programs as a long-term approach to end poverty, improve nutrition, and sustain economic growth. These crops are well-suited for commercial, small scale, and subsistence farming and actually constitute over 90% of the products grown in home gardens.

Currently, pests such as insects, mites, bacteria, fungi, viruses, nematodes, and weeds often cause up to 40% reduction in crop yields (SP-IPM, 2010). Insect pests are responsible for a major part of these losses due to direct feeding damage and/or vectoring of viral, bacterial, and fungal diseases. Recognizing the need to minimize this loss, the U.S. Agency for International Development (USAID) and other international donors, such as the United Nations Food and Agriculture Organization (FAO) and the World Bank, have been supporting local and national governments, NGOs, and farmers through the establishment of integrated pest management programs. In Indonesia, the National IPM Program was established in the mid-1980s with support from USAID and FAO. This program utilized Farmer Field Schools (FFS) as the main conduit for introducing IPM concepts to farmers. Due to its success, the FFS concept has spread to many other countries. The USAID-funded Integrated Pest Management Collaborative Research Support Program (IPM CRSP) began supporting IPM activities in Asia and other countries in 1993 and still supports these activities to date.

Arthropod Pests of Horticultural Crops in Tropical Asia (R. Muniappan *et al.*)

Development of effective pest management programs cannot move forward unless the pests are correctly identified and their distribution, biology, nature of damage, and important natural enemies are known. This book summarizes available information on insect pests and their natural enemies on horticultural crops in tropical Asia. It also builds on the publication of Shepard *et al.* (1999) in that it contains an abundance of color photographs that help to identify insect pests and their natural enemies. In addition, this current work expands into fruit and specialty crops.

Forty-three horticultural crops are covered, of which 15 are vegetables, 24 are fruit tree crops, and 4 are in other categories that include cashew, coffee, cocoa, and tea. These crops, originated from many parts of the Old and New World and are cultivated in lowlands as well as higher elevations. Several tropical Asian countries have elevations where temperate vegetable crops, such as carrot, potato, and crucifers (broccoli, cauliflower, cabbage, etc.), as well as beverage crops like coffee and tea, are grown. Litchi, durian, mangosteen, and rambutan are subtropical trees, but grow well in the Asian tropics

Understanding the center of origin for both the crops and their pests can be important for developing effective pest management programs. This book provides information on the center of origin of crops and their insect pests, where this information is available. As expected, the crops that originated in Asia have more arthropods associated with them than crops introduced from the neotropical, European, and Afro-tropical regions. This is especially true for invasive pests that originated in one region and were accidentally introduced into another without their natural enemies to keep them in check. Understanding the origin is especially important for invasive pests, such as the papaya mealybug *Paracoccus marginatus* Williams and Granara de Willink (Hemiptera: Pseudococcidae). It spread from Mexico in the early 1990s, but parasitoids were identified in its center of origin and were introduced to several countries in the Caribbean, South America, Pacific Asia and Africa where they effectively controlled this pest (Muniappan *et al.*, 2009, 2010). These parasitoid introductions resulted in billions of dollars in savings and averted the collapse of the papaya industry.

Another example is the eggplant fruit and shoot borer *Leucinodes orbonalis* Guenée (Lepidoptera: Pyralidae). This pest is a native of India, but has spread to South and Southeast Asia and Africa (Waterhouse, 1998). It became established in the Philippines relatively recently (Navasero, 1983) and has become one of the most serious pests of eggplant in that country. Even though several natural enemies have been reported for this pest in India (Yasodha and Natarajan, 2006), the high levels of damage exhibited by local eggplant cultivars indicate that these natural enemies are not effective. Several parasitoids have been reported on a closely related insect, *Neoleucinodes elegantalis* (Guenée) (Lepidoptera: Pyralidae), a pest of eggplant and tomato in the Neotropics, which may have potential for biological control of *L. orbonalis* (Waterhouse, 1998). These parasitoids provide an opportunity to test the hypothesis of Hokkanen and Pimentel (1984) that the success of introducing natural enemies for biological control is about 75% higher if new parasitoid-host associations are employed rather than those based on long-evolved associations between parasitoids and hosts. Information on natural enemies presented in this book should provide useful information for researchers who are searching for biocontrol agents.

Understanding the ecology of the pests and their natural enemies plays a major role in developing IPM strategies (Peña *et al.*, 2002). Crucifer crops in the tropical regions of Asia are attacked by a dozen or more insect pests. Upland areas have a different pest complex than the lowlands. *Plutella xylostella* (L.) (Lepidoptera: Plutellidae) and *Crocidolomia pavonana* (F.) (Lepidoptera: Pyralidae) are present in the higher elevations, while *P. xylostella* and *Hellula undalis* (F.) (Lepidoptera: Pyralidae) occur in lower elevations of the tropics in Asia. The parasitoids, *Cotesia plutellae* Kurdjumov (Hymenoptera: Braconidae) in the lowlands and *Diadegma semi clausum* (Hellen) (Hymenoptera: Ichneumonidae) in the uplands have proven effective for controlling *P. xylostella*. Other generalist predators and entomopathogens also have been found to play a major role in suppression of crucifer pests. We found epizootics of *Zoophthora radicans* in populations of *P. xylostella* in North Sulawesi, Indonesia; when larvae were collected and held in the laboratory, 100% of them came down with the disease. Thus, if the farmer had knowledge of the benefits of this natural enemy, chemical sprays and their attendant negative side effects could have been avoided altogether.

Viral diseases of vegetable crops transmitted by aphids, leafhoppers, mealybugs, thrips, whiteflies, and scale insects are serious problems in the tropics and require management of both insect vectors and viruses through an integrated approach. This book provides information on important insect vectors and the viruses that they transmit; this is important for developing effective management programs. Managing these viral diseases involves integration of several tactics such as resistant varieties, use of natural enemies, roguing diseased plants, establishing host free periods, and screening seedbeds (IPM CRSP, 2009, 2010).

We hope that the information on the natural history of arthropod pests and their natural enemies, along with associated color photographs, will help to build a solid foundation for sustainable IPM programs. This book should be useful to students, researchers, extensionists, farmers, and others who are interested in management of arthropod pests of horticultural crops in tropical Asia.

2 Pests of Major Vegetable Crops

2.1 Pests of Beans (*Phaseolus* spp., *Vigna* spp., and others, Fabaceae)

The common name, beans, is used to refer to plants belonging to several genera in the family Fabaceae (Leguminosae). Beans are rich in protein and high in the amino acids lysine and methionine, making beans nutritionally complementary to cereals, which are deficient in these two amino acids. Beans are the main source of protein in a vegetarian diet and are featured predominantly in Asian traditional cuisine. We primarily address arthropod pests which infest beans used as vegetables. Although several pests attack beans, the bean pod borer (*Maruca vitrata*) and aphids (mostly *Aphis craccivora*) are the most important. The bean pod borer causes direct damage by boring into pods; aphids cause direct damage by sucking plant juices and secreting honeydew that causes the build-up of sooty mold that blocks photosynthesis. Aphids also cause indirect damage by transmitting important plant viruses.

Common name: Bean pod borer
Scientific name: *Maruca vitrata* (F.)
Order: Lepidoptera
Family: Crambidae
Distribution: Pantropical, Africa, Australia, tropical Asia, North and South America, and Pacific Islands
Hosts: Legumes
Life cycle: Adult females lay eggs singly or in groups of two to ten on flowers, flower buds, or pods. Eggs are oval in shape and greenish white when laid, later turning to pale yellow. The incubation period ranges from 2 to 4 days. Newly hatched larvae feed on flowers and later bore into the bean pods. Larval duration is about 10 days. Pupation takes place in a silken cocoon in the pod or in dried leaves on the ground (Figs 2.1.1, 2.1.2, 2.1.3).

Fig. 2.1.1. *Maruca vitrata* adult

Fig. 2.1.2. Frass on pods due to *Maruca vitrata* larval feeding

Fig. 2.1.3. *Maruca vitrata* larva

Damage: This pest causes heavy losses in beans by feeding on flower buds, flowers, pods, and terminal shoots.
Natural enemies: Waterhouse and Norris (1987) reported natural enemies of *M. vitrata* that occur in different countries as well as those introduced for biological control. The parasitoids, *Antrocephalus* nr. *subelongatus, Bracon greeni, Phanerotoma hendecasisella, Microdus* sp., *Ceraphron* sp., *Nesolynx thymus,* and *Telenomus* sp. from Sri Lanka; *Apanteles antipoda* from West Indies; *Apanteles etiellae isolatus, Bracon cajani, Bracon hebetor, Bracon kirkpatricki, Bracon pectoralis, Bracon piger, Bracon thurberiphagae, Eiphosoma dentator, Goniozus* (= *Perisierola*) sp., *Phanerotoma bennetti, Chelonus busckiella, Chelonus inanitus,* and *Parasierola emigrata* from Trinidad; *Macrocentrus ancylivora* from USA; *Phanerotoma planifrons* from France; *Bracon cajani* from Mauritius; *Phanerotoma leucobasis, Braunsia kriegeri, Bassus bruesi, Pristomerus* sp., *Dolichogenidea* sp., *Testudobracon* sp., and *Cadurcia* sp. from Benin; *Cremnops* sp. and *Trichogramma evanescens* from the Philippines; *Baeognatha* (= *Agathis*) sp. and *Cotesia* sp. from Indonesia; and *Cedria paradoxa* and *Bracon hebetor* from India were recorded. The parasitoids, *Phanerotoma philippinensis* and *Bassus javanicus* were observed in Southeast Asia (Tamò *et al.*, 2003). *Apanteles* sp., *Microbracon thurberiphagae, Brachymeria ovata, Nemorilla floralis,* and *Argyrophylax* (*Sturmia*) *albincisa* have been recorded in Central America (King and Saunders, 1984). Sharma (1998) listed parasitoids and predators reported from various countries. A parasitoid, *Apanteles taragamae,* identified as the most promising candidate by the Asian Vegetable Research and Development Center, was introduced from Taiwan into Benin, Ghana, and Nigeria in 2007. However, its establishment has not been confirmed. A nucleopolyhedrovirus (MaviMNPV) found in Taiwan was introduced to the Mono region of Benin (IITA, 2010).

Common names: Bean lycaenid, gram blue
Scientific name: *Euchrysops cnejus* (F.)
Order: Lepidoptera
Family: Lycaenidae
Distribution: Asia, Australia
Hosts: Beans
Life cycle: Eggs are spherical in shape and are laid singly on young beans or flower buds. The larval stages last from 11 to 14 days. Larvae grow to 18 mm in length. The pupal stage lasts 6–7 days (Figs 2.1.4, 2.1.5).
Damage: Larvae feed on pods of longbean and other beans. Damage is characterized by round holes in pods and feeding on pod contents.
Natural enemies: Several species of parasitoids have been recorded.
Other remarks: Often, the larvae are found in association with the ant *Camponotus compressus.*

Common names: Pea blue butterfly, long-tailed blue, bean butterfly
Scientific name: *Lampides boeticus* (L.)
Order: Lepidoptera
Family: Lycaenidae
Distribution: Asia, Africa, and Oceania
Hosts: Legumes
Life cycle and Damage: Similar to *E. cnejus.*

Fig. 2.1.4. *Euchrysops cnejus* adult butterfly

Fig. 2.1.5. *Euchrysops cnejus* larva near the bore hole on a bean pod

Common name: Bean fly
Scientific name: *Ophiomyia phaseoli* (Tryon) (= *Agromyza phaseoli, Melanagromyza phaseoli*)
Order: Diptera
Family: Agromyzidae
Distribution: Africa, Asia, Australia, and Pacific Islands
Hosts: *Phaseolus vulgaris, Vigna* spp., *Cajanus cajan, Glycine max, Lablab niger, Pisum sativum,* other wild beans
Life cycle: This small, shiny black fly lays eggs under the epidermis of the upper surface of the young leaves. It lays about 100–300 eggs during its lifespan of two weeks. The egg stage lasts 2–4 days. Larvae pass through three instars, and the larval stage is completed in about 10 days at warmer temperatures. In cooler areas, development time is extended to 17–20 days. Larval feeding and pupation occur mainly in the center of the stem near the soil surface, but in older plants pupation occurs at the junction of the leaf and stem. The pupal stage lasts 7–13 days in colder areas (Fig. 2.1.6).
Damage: This pest can be a major constraint to legume production. The main damage is caused by the larvae as they mine the central core of the stem down to the root-shoot junction. When attacks occur during the seedling stage, plants normally wilt and die (Figs 2.1.7 and 2.1.8).
Natural enemies: More than 50 parasitoids have been recorded from the bean fly. Parasitoids reported from Asia include *Fopius* sp., *Plutarchia indefensa,* and *Euderus* sp. from Thailand; *Eurytoma poloni* and *Sphegigaster voltairei* from Indonesia; *E. poloni,*

Fig. 2.1.6. *Ophiomyia phaseoli* adult fly

Fig. 2.1.7. Seedling killed by *Ophiomyia phaseoli* infestation

Fig. 2.1.8. Bean leaf petiole damaged by *Ophiomyia phaseoli*

Plutarchia sp., and *Sphegigaster stella* from Malaysia; *S. stella, Opius phaseoli*, and *Hemiptarsenus* sp. from the Philippines; *Opius phaseoli, Tetrastichus* sp., *Eurytoma* sp., *Chlorocytus* sp., *Polycystus* sp., *Sphegigaster brunneicornis*, and *Sphegigaster rugosa* from India; and *Polycystus propinquus* and *S. rugosa* from Sri Lanka. *Opius phaseoli* and *Opius importatus* from Uganda were introduced to Hawaii and *O. phaseoli* to Guam and Taiwan from Hawaii for biological control of the bean fly (Waterhouse, 1998).

Common name: Cowpea aphid
Scientific name: *Aphis craccivora* Koch
Order: Hemiptera
Family: Aphididae
Distribution: Cosmopolitan
Hosts: This 1.4–1.9 mm long insect is polyphagous. Some of the common hosts are cowpea, pigeon pea, beans, citrus, okra, coffee, peanut, and cocoa.
Life cycle: Under favorable climatic conditions, wingless, parthenogenetic females produce nymphs viviparously, and, when conditions are not favorable, winged parthenogenetic females are produced. In temperate areas, winged males and sexual, egg-laying females are produced. Nymphs pass through four molts to become adults in 5–8 days (Fig. 2.1.9).
Damage: This aphid feeds on stems and terminal shoots of seedlings, and on flowers and pods of mature plants. Heavy infestations can kill seedlings, and can cause flower drop and pod shriveling in mature plants. Besides direct removal of plant sap, this aphid is known to transmit about 30 plant viruses, including viruses of legumes, cucurbits, and crucifers. On citrus, it transmits tristeza virus. Indirect damage from the build-up of sooty mold, which grows on the honey dew secretions, also can affect plant growth.

Fig. 2.1.9. *Aphis craccivora* infesting a bean pod

Natural enemies: The parasitoids, *Lipolexis scutellaris* in Malaysia and the Philippines; *Aphelinus abdominalis* in Pakistan and India; *Aphidius absinthii, Aphidius matricariae, Ephedrus* sp. nr. *cerasicola, Lysiphlebus fabarum, Lysiphlebus* sp., and *Trioxys sinensis* in Pakistan; and *Diaeretiella rapae, Ephedrus persicae, Lipolexis scutellaris, Lysaphidus* sp., *Lysiphlebus testaceipes, Praon abjectum, Toxares zakai,* and *Trioxys indicus* in India have been recorded. Most predatory coccinellids feed on aphids. Predators recorded in India and Pakistan include *Anthocoris* spp., *Orius* sp., *Chrysoperla carnea, Adonia variegata, Brumus suturalis, Cheilomenes septempunctata, Harmonia dimidiata, Pullus* spp., *Scymnus* spp., *Leucopis* sp., *Episyrphus balteatus, Eupeodes confrater, Ischiodon scutellaris, Microdon bicolor, Paragus tibialis, Paragus serratus,* and *Syrphus* spp. A beneficial fungus, an Entomophthorales, has been found infecting this aphid in Indonesia. Parasitoids, *Aphidius* spp. and predators, *Ceratomegilla maculata, Cycloneda sanguineda, Hippodomia convergens, Allograpta obliqua,* and *Chrysopa* spp. were recorded in Central America (King and Saunders, 1984).

Common names: Green vegetable bug, southern green stink bug
Scientific name: *Nezara viridula* (L.)
Order: Hemiptera
Family: Pentatomidae
Distribution: This pest occurs throughout tropical and subtropical regions.
Hosts: Legumes, tomato, capsicum, guava, others
Life cycle: The barrel-shaped eggs are laid in clusters of 10–90 on the undersides of leaves. The eggs turn reddish a few days before the nymphs hatch. The egg stage lasts 5–7 days. Newly hatched nymphs remain near the eggshells for a day or two before dispersing. The brightly colored nymphs pass through five instars in 21–28 days. Nymphs are multicolored, but the adults are green. Adults live up to 3 weeks (Figs 2.1.10, 2.1.11, 2.1.12).
Damage: A wide range of crops, especially legumes, are attacked by this bug. Fruits and fruiting bodies are normally the target for feeding. Bugs may also carry spores of fungal pathogens from plant to plant and mechanically transmit plant pathogens while feeding.
Natural enemies: About 80 parasitoids recorded on this bug were listed by Waterhouse (1998). Three species of egg parasitoids are known from Indonesia, namely: *Ooencyrtus malayensis, Trissolcus basalis,* and *Telenomus cyrus*. Parasitoids reported from other Southeast Asian countries are *Trissolcus rudus* and *Telenomus* sp. from Vietnam; *Telenomus pacificus, T. cyrus, Telenomus comperei, O. malayensis* from the Philippines; and *O. malayensis* from Malaysia. *Xenoencyrtus niger* from Australia; *Anastatus* sp. from Pakistan; the tachinid fly *Trichopoda pilipes* from Florida and West Indies; *Ooencyrtus submetallicus* and *Ooencyrtus trinidadensis* from West Indies; and *Trissolcus mitsukurii* and *Trissolcus nakagawai* from Japan were also recorded (Waterhouse and Norris, 1987). The parasitoids *Telenomus* spp. and *Trichopoda pennipes* from Central America were recorded (King and Saunders, 1984). Other natural enemies, particularly predators, impact this bug and help to keep its populations low. Eggs and first instar nymphs are favored by generalist predators. In Malaysia, the predaceous reduviid bug *Sycanus collaris* keeps this pest under control.

Fig. 2.1.10. *Nezara viridula* adult

Fig. 2.1.11. *Nezara viridula* eggs

Fig. 2.1.12. *Nezara viridula* nymphs

Common name: Legume shield bug
Scientific name: *Piezodorus hybneri* Gmelin
Order: Hemiptera
Family: Pentatomidae
Distribution: Southeast Asia
Hosts: Legumes, polyphagous
Life cycle: Eggs are laid in clusters of two rows on leaves, stems, or pods. Each egg cluster contains 9–42 eggs. Initially, nymphs stay together after hatching, but disperse as they mature. There are five nymphal stages lasting 14–22 days. Adults are about 10 mm long (Figs 2.1.13, 2.1.14, 2.1.15).

Fig. 2.1.13. *Piezodorus hybneri* adult

Fig. 2.1.14. Egg mass of *Piezodorus hybneri*

Fig. 2.1.15. Nymphs of *Piezodorus hybneri*

Damage: Both adults and nymphs feed on pods of legumes, causing losses in fruit quality.

Natural enemies: The egg parasitoid *Telenomus triptus* was reported in Japan. Other egg parasitoids and predators are important control factors.

Fig. 2.1.16. *Chrysodeixis chalcites* adult

Fig. 2.1.17. Larva of *Chrysodeixis chalcites*

Common name: Garden looper
Scientific name: *Chrysodeixis* (*Plusia*) *chalcites* (Esper)
Order: Lepidoptera
Family: Noctuidae
Distribution: Southeast Asia
Hosts: Beans, cucurbits, crucifers, carrots, and legumes
Life cycle: Eggs are laid singly on the leaves and hatch in 3–4 days. Each female lays about 200 eggs. There are five larval instars, and the larval period lasts for about 13 days. Young larvae feed on the under surface of leaves, leaving the upper cuticle intact. Later instars consume whole pieces of leaves. Pupation takes place in a silken cocoon within a folded leaf. Adults emerge after 7 days (Figs 2.1.16, 2.1.17).
Damage: Larvae feed on leaves and only occasionally cause defoliation serious enough to reduce yields.
Natural enemies: In Indonesia, many egg and larval parasitoids such as Copidosomopsis sp. *Glyptapanteles phytometrae* and *Cotesia* sp. have been reared from field-collected larvae. In addition, there are carabid and pentatomid predators as well as wasps that attack this pest. Insect pathogens play a major role in the regulation of *C. chalcites* populations, notably the fungi *Pandora gammae* and *Nomuraea rileyi*, and a nucleopolyhedrovirus. Because of the combined action of these natural enemies, *C. chalcites* rarely causes economic damage (Figs 2.1.18, 2.1.19).

Common name: Nettle caterpillar
Scientific name: *Parasa philepida* (*lepida*) (Cramer)
Order: Lepidoptera
Family: Limacodidae
Distribution: Asia

Fig. 2.1.18. Parasitoid *Glyptapanteles* pupa near *Chrysodeixis* larva

Fig. 2.1.19. *Chrysodeixis* parasitized by *Copidosomopsis* sp.

Hosts: Banana, beans, rambutan, coffee, coconut
Life cycle: Each female lays 400–600 eggs in batches that hatch in 3–5 days. Young larvae feed on leaf epidermis and, as they grow, consume entire leaves. There are seven larval stages (often eight for females) which are completed in 35–42 days. Pupation occurs in cocoons often attached to the stem or bark, and the pupal stage lasts 21–24 days. The pupa may lie dormant during the dry season. Adults emerge with the onset of rain (Fig. 2.1.20).
Damage: Caterpillars of this pest cause defoliation when they occur in large numbers.
Natural enemies: Larval parasitoids *Apanteles parasae* and *Praestochrysis shanghaiensis*; a fungal pathogen, *Cordyceps coccinea*; and virus diseases keep this pest under control in Indonesia.

Other major pests are:
Corn earworm, *Helicoverpa armigera* (See Pests of Tomato)
Broad mite, *Polyphagotarsonemus latus* (See Pests of Pepper)
Spider mites, *Tetranychus* spp. (See Pests of Eggplant)
Leafminers, *Liriomyza* spp. (See Pests of Tomato)
Cluster caterpillar or army worm, *Spodoptera litura* (See Pests of Tomato)
Sweetpotato whitefly, *Bemisia tabaci* (See Pests of Okra)

Some regionally important pests are the pod bugs *Clavigralla gibbosa* Spinola and *Riptortus pedestris* (Fabricius) (Hemiptera: Coreidae); the plume moth *Exelastis atomosa* (Walsingham) (Lepidoptera: Pterophoridae); the field-bean pod borer *Adisura atkinsoni* Moore (Lepidoptera: Noctuidae); pod borer *Etiella zinckenella* Treitschke (Lepidoptera: Pyralidae); the pod fly *Melanagromyza obtusa* (Malloch) (Diptera: Agromyzidae); and the blister beetle *Mylabris pustulata* (Thunberg) (Coleoptera: Meloidae).

Fig. 2.1.20. Larvae and pupa of *Parasa philepida*

2.2 Pests of Cabbage and other Crucifers (Cruciferae)

Cruciferous vegetables are economically important throughout the world. They are normally considered temperate crops, but also are an important component of the diet in tropical areas. They are a valuable source of nutrition, including vitamins A, C, niacin, folic acid, and dietary fiber. In tropical Asia, crucifers are grown by small landholders in specialized production areas, such as around urban centers and in upland regions. Sales of these vegetables provide a valuable source of income, so farmers often use input-intensive agronomic practices, particularly the heavy use of insecticides. In recent years, crucifer production has been seriously affected by a steady increase in insect pests because of heavy pesticide use and insect resistance to these chemicals. Crucifers grown in upland areas often have different key insect pests. In higher elevations the diamondback moth, *Plutella xylostella*, and the cabbage cluster caterpillar, *Crocidolomia pavonana* are usually the most important pests, while in lowland areas, the cabbage webworm, *Hellula undalis* and the diamondback moth are the two major pests.

Common name: Diamondback moth
Scientific name: *Plutella xylostella* (Linnaeus)
Order: Lepidoptera
Family: Plutellidae
Distribution: Cosmopolitan
Hosts: Crucifers
Life cycle: Eggs are oval, flat, and laid singly or in groups up to eight, mostly on the upper side of the leaves. The incubation period is 4–8 days. Upon hatching, the first

instar larva crawls to the lower surface of the leaf, bores into the epidermis, and mines into the tissue. Subsequent instars feed on the lower surface of the leaf, leaving the upper epidermal layer intact, and resulting in a characteristic window effect. Larvae are green, and the larval duration ranges from 9 to 30 days. Pupation takes place inside a finely woven silken cocoon, and the pupal duration ranges from 5 to 15 days (Figs 2.2.1, 2.2.2, 2.2.3).

Damage: Fourth instar larvae cause most of the feeding damage. Unless proper control measures are taken, they are capable of totally destroying the crop.

Natural enemies: Eggs are parasitized by *Trichogramma japonicum* and *Trichogramma chilonis* and the larvae by *Diadegma semiclausum* (Fig. 2.2.4), *Diadromus collaris*, *Cotesia plutellae* (Fig. 2.2.5), and *Ooencyrtus* (*Tetrastichus*) *sokolowskii*. In 1931, *Diadegma fenestralis* was introduced from the Netherlands to Indonesia, but it failed to establish. In 1950, *Diadegma eucerophaga* was introduced from New Zealand to Indonesia, and by 1952 it was causing 72–96% parasitism of the diamondback moth. The parasitoids *D. eucerophaga* from Indonesia and New Zealand; *D. collaris* from Australia, India, and New Zealand; and *O* (*T.*) *sokolowskii* from India were introduced to Malaysia in the mid 1970s and established. Another parasitoid, *Macromalon orientale*, introduced from India to Malaysia, did not establish. In the

Fig. 2.2.1. Adult of *Plutella xylostella*

Fig. 2.2.2. Larva of *Plutella xylostella*

Fig. 2.2.3. Pupa of *Plutella xylostella*

Fig. 2.2.4. Parasitoid *Diadegma semiclausum*

Fig. 2.2.5. Parasitoid *Cotesia plutellae*

Fig. 2.2.6. Fungus *Hirsutella* sp. infecting *Plutella xylostella* larva

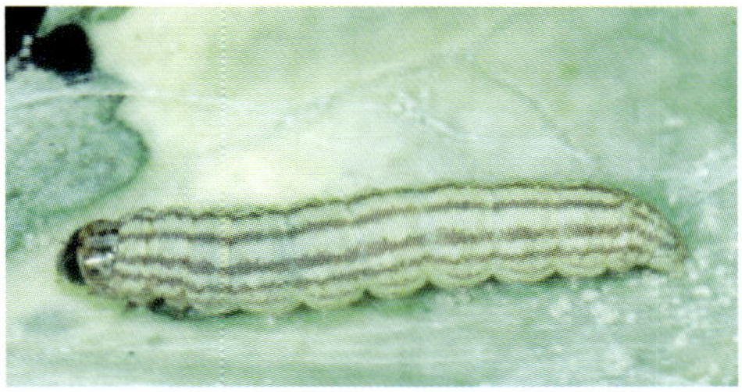

Fig. 2.2.7. Larva of *Hellula undalis*

Philippines, *C. plutellae* is attacked by a hyperparasitoid. Other important natural enemies include the fungi *Zoophthora radicans* (Brefeld) Batko and *Hirsutella* sp. (Fig. 2.2.6).

Common name: Cabbage webworm
Scientific name: *Hellula undalis* (F.)
Order: Lepidoptera
Family: Pyralidae
Distribution: Europe, Africa, Asia, Pacific Islands
Hosts: Brassicaceae
Life cycle: Eggs are ovoid and white when laid and turn pinkish the next day. Eggs are laid singly or in groups on the leaves and hatch in 2–9 days. Newly-hatched larvae bore into the leaf tissue and either tunnel into the leaf lamina and midrib or graze on the plant tissue while remaining under a silken web. Larvae often feed on the terminal buds, causing distortion and death of seedlings. There are five instars (Fig. 2.2.7) and the larval stage ranges from 16 to 19 days. Pupation takes place in the feeding tunnel or in the soil in a silken cocoon. The pupal stage lasts 6–10 days.

Damage: This pest causes serious loss at the seedling stage by boring into the growing tips. Also, the boring of larvae ruins cabbage and cauliflower heads. It is an important pest in tropical lowlands, but not at higher elevations, as the lower temperatures affect its development.
Natural enemies: Polyphagous parasitoids have been reported, but no specific ones have been identified. The parasitoids *Apanteles* sp., *Agathis* sp., and *Diadegma* sp. and a fungus *Hirsutella* sp. have been recorded from Egypt. Trap cropping with radish in cabbage fields was recommended in Guam.

Common name: Cabbage cluster caterpillar
Scientific name: *Crocidolomia pavonana* (F.)
Order: Lepidoptera
Family: Pyralidae
Distribution: Africa, Asia
Hosts: Brassicaceae
Life cycle: Eggs are laid in clusters of 8–114, overlapping each other on the lower surface of the leaves. They hatch in 4–5 days. Newly-hatched larvae remain near the egg mass and feed on the leaf surface. As they grow, the larvae disperse throughout the plant. A cabbage plant can be destroyed by larvae from a single egg mass. In some cases, they reach the heart of the cabbage, riddle the plant tissue, and produce mass webbing and frass. The larval stage lasts about 12 days and there are five instars. Pupation takes place in a silken cocoon in the soil. The pupal stage lasts about 10 days (Figs 2.2.8, 2.2.9, 2.2.10, 2.2.11).
Damage: This is one of the most important pests of cole crops. It can destroy a crop of cabbage or cauliflower unless timely control measures are taken.
Natural enemies: The parasitoids *Trathala flavoorbitalis* from the Philippines; *Apanteles crocidolomiae*, *Apanteles obliquae*, *Bracon lefroyi*, *Bracon hebetor*, *Bracon melleus*, *Diplazon orientalis*, *Palexorista solennis*, and *Enicospilus xanthocephalus* from India; *Diadegma* sp., *Exochus* sp., *Palexorista inconspicuoides*, and *P. solennis* from Papua New Guinea; *Colpotrichia* sp., *Diadegma* sp., *Eriborus argenteopilosus*, and *Bleparipa* (= *Sturmia*) sp. from Indonesia (Shepard and Barrion, 1998) have been recorded on *C. pavonana.* A predatory bug, *Eocanthecona furcellata*, was also reported from India.

Fig. 2.2.8. Adult of *Crocidolomia pavonana*

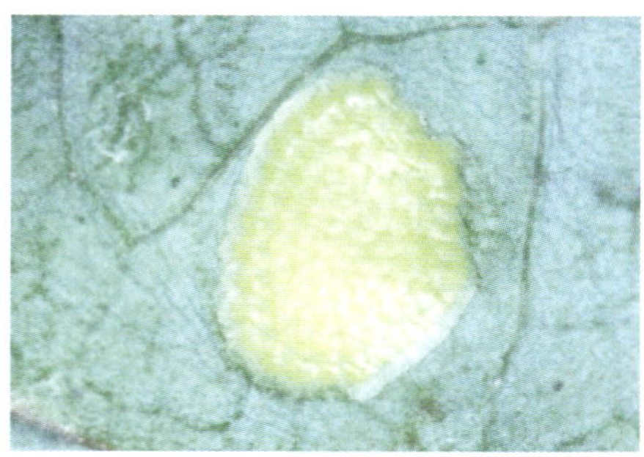

Fig. 2.2.9. Egg mass of *Crocidolomia pavonana*

Fig. 2.2.10. Larvae of *Crocidolomia pavonana*

Fig. 2.2.11. Cabbage plant damaged by *Crocidolomia pavonana*

Common names: Imported cabbageworm, the cabbage white
Scientific name: *Pieris rapae* (L.)
Order: Lepidoptera
Family: Pieridae
Distribution: Asia, North Africa, Europe, Australia, North America
Hosts: Crucifers
Life cycle: Adults deposit single eggs that hatch in 3–4 days depending upon temperature. Larvae undergo five developmental stages, and pupation takes place on the plant. Mature larvae can reach a length of about 30 mm (Figs 2.2.12, 2.2.13, 2.2.14).
Damage: Larval feeding on leaves of crucifers can cause serious damage.
Natural enemies: Little is known about natural enemies of this insect in Asia. In India, *Apanteles glomeratus* (Lineaeus) has been reported. In some parts of the world, larval and pupal parasitoids (especially *Apanteles* spp. and *Cotesia* spp.) and a granulosis virus are important natural control factors.

Fig. 2.2.12. Adult of *Pieris rapae*

Fig. 2.2.13. Larva of *Pieris rapae*

Fig. 2.2.14. Chrysalis of *Pieris rapae*

Other remarks: Other species of *Pieris* that occur in Asia are *P. brassicae* (L.), and *P. canidia* (Sparrman). Life cycles and damage are similar to those described for *P. rapae* (above).

Common name: Cabbage shield bug, harlequin bug
Scientific name: *Eurydema pulchrum* (Westwood)
Order: Lepidoptera
Family: Pentatomidae
Distribution: South and Southeast Asia, Australia
Hosts: Brassicas, radish
Life cycle: Eggs are laid in batches of 12, lined in two rows. Eggs are cylindrical in shape and the egg stage lasts about 6 days. Upon hatching, nymphs feed gregariously, then they gradually disperse as they grow older. There are five instars lasting for about 28 days. Adults (7 mm) are colorful with orange, black, and white markings (Fig. 2.2.15).
Damage: Feeding by nymphs and adults causes white necrotic spots on the leaves of crucifer crops. Damage is localized and confined to the older, outer leaves.
Natural enemies: Little is known about the natural enemies of this pest.

Fig. 2.2.15. Adult of *Eurydema pulchrum*

Common name: Painted bug
Scientific name: *Bagrada cruciferarum* Kirkaldy
Order: Hemiptera
Family: Pentatomidae
Distribution: India, Myanmar, Sri Lanka, Iraq
Hosts: Crucifer crops
Life cycle: Adults are sub-ovate in shape with black and orange spots. The adult female lays oval, pale yellow eggs singly or in groups on leaves and stems. The number of eggs laid varies from 37 to 102, and eggs hatch in 3–5 days. There are five nymphal instars and their developmental period ranges from 16–22 days.
Damage: Both nymphs and adults feed on leaves and developing pods by sucking the sap.
Natural enemies: Eggs are parasitized by *Gryon* sp. and adults by *Alaphora* sp.

Common name: Flea beetle
Scientific name: *Phyllotreta* sp.
Order: Coleoptera
Family: Chrysomelidae
Distribution: Cosmopolitan
Hosts: Polyphagous
Life cycle: Eggs are laid in the soil, and the larvae feed on roots. Pupation takes place in small cocoons in the soil. Total life cycle from egg to adult is about 3–4 weeks. Adult beetles are 2 mm long (Fig. 2.2.16).
Damage: Adults feed on leaves, scraping leaf tissue and making characteristic perforations. When large populations occur, these perforations make the vegetable unmarketable.
Natural enemies: None reported.

Fig. 2.2.16. Adult beetle of *Phyllotreta* sp.

Common name: Mustard sawfly
Scientific name: *Athalia proxima* (Klug)
Order: Hymenoptera
Family: Tenthredinidae
Distribution: India, China, Indonesia, Myanmar, Taiwan
Hosts: Mustard, cabbage, radish, cauliflower, turnip
Life cycle: The adult is a shiny black wasp with yellow thorax and femora. Eggs are inserted singly along the leaf margins. They are cream colored and hatch in 3–4 days. Larvae are dark gray and have three pairs of thoracic and eight pairs of abdominal legs. They pass through five to seven instars and the larval duration varies from 13–35 days depending upon the temperature (Fig. 2.2.17). Pupation takes place in an oval cocoon in the soil.
Damage: Larvae feed on the leaves and skeletonize them.
Natural enemies: The larval parasitoid, *Perilissus cingulator*, and a bacterium *Serratia marcescens* have been recorded from India.

Common name: Cabbage aphid
Scientific name: *Brevicoryne brassicae* (L.)
Order: Hemiptera
Family: Aphididae
Distribution: Cosmopolitan, but mostly in higher altitudes in the tropics
Hosts: Crucifer crops
Life cycle: The body of this aphid is covered with a white waxy layer. Both winged and apterous forms are found in the field. Females produce live young. Males are produced when there is a decrease in temperature or photoperiod. Honeydew produced by this aphid induces growth of sooty mold on the leaves (Hill, 1987).
Damage: Feeding by this aphid results in stunted growth and yellowing of leaves. Leaves become sticky due to honeydew deposition and black because of sooty mold.
Natural enemies: The parasitoid, *Diaeretiella rapae*, syrphid fly maggots, lacewing larvae, and ladybeetles commonly feed on this aphid.

Fig. 2.2.17. Larva of *Athalia* sp.

Common name: False cabbage aphid
Scientific name: *Lipaphis erysimi* (Kalt.)
Order: Hemiptera
Family: Aphididae
Distribution: Tropical and subtropical regions
Hosts: Crucifer crops, bean, beet, onion, cucumber, potato, others
Life cycle: This yellowish green aphid multiplies parthenogenetically and both apterous and winged forms are known to occur. There are four nymphal instars. Adults live 13–15 days and deposit about 100–200 nymphs.
Damage: Feeding by this aphid stunts the growth of the plant, and honeydew produced induces sooty mold which reduces the photosynthetic surface of leaves. *Liphaphis erysimi* is a vector of viral diseases like *Cabbage black ring spot virus, Cauliflower mosaic virus, Celery mosaic virus, Bean common mosaic virus, Radish mosaic virus, Cucumber mosaic virus,* and others.
Natural enemies: Ladybeetles, *Coccinella septempunctata, Adalia bipunctata* (L.), *Cheilomenes vicina,* syrphids (Fig. 2.2.18), *Xanthogramma scutellare,* and *Betasyrphus adligatus* have been recorded on this aphid. A parasitoid, *Diaeretiella rapae,* and the fungi *Cephalosporium aphidicola* and *Pandora neoaphidis* were also reported on this aphid.

Other major pests are:
Tomato fruit worm, *Helicoverpa armigera* (See Pests of Tomato)
Cluster caterpillar, army worm, *Spodoptera litura* (See Pests of Tomato)
Leaf miners, *Liriomyza* spp. (See Pests of Tomato)
Garden looper, *Chrysodeixis chalcites* (See Pests of Beans)
Green peach aphid, *Myzus persicae* (See Pests of Pepper)

Fig. 2.2.18. Syrphid maggot feeding on an aphid

2.3 Pests of Cassava (*Manihot esculenta* Crantz, Euphorbiaceae)

Common name: Cassava mealybug
Scientific name: *Phenacoccus manihoti* Matile-Ferrero
Order: Hemiptera
Family: Pseudococcidae
Distribution: South America, Africa, Thailand, Indonesia, Cambodia
Hosts: This pest is oligophagous and feeds on plants in nine families but it prefers cassava.
Life cycle: Eggs are yellowish-gold and oblong and are contained in ovisacs at the end of the females' abdomen. Although females are thought to give live birth, eggs hatch soon after they are laid. The life cycle consists of an egg stage and four instars, the latter being the adult. Development rate is temperature dependent but the average adult life span for females is 20 days. The egg stage lasts about 8 days and the adult female dies within 1–3 days after egg production ceases. She produces about 500 eggs during her life span. Crawlers (early-stage nymphs) are light yellow and are flattened with smooth bodies. As soon as the nymphs hatch, they emit a white, waxy coating over their bodies, giving them a "mealy" appearance. The complete life cycle (egg to adult) is about 33 days. Females are covered with white waxy filaments. Only the males have wings and they only live for a few days.
Damage: This mealybug is specific to cassava and is not a problem in South America, its center of origin. It was first observed outside South America in Zaire in 1973, causing severe distortion of leaves. It spread into almost all other cassava-growing areas in Africa by 1982 (Kogan *et al.*, 1999; Neuenschwander *et al.*, 2003). The losses caused by this mealybug, together with another introduced pest, cassava green mite, were estimated to be $2 billion per year (Herren and Neuenschwander, 1991) in Africa. Cassava mealybug was accidentally introduced to Thailand (2009) and Cambodia and Indonesia (2010) (Fig. 2.3.1).
Natural enemies: Extensive surveys conducted in South America identified the parasitoids *Aenasius vexans*, *Apoanagyrus diversicornis*, *Anagyrus* spp., and *Apoanagyrus*

Fig. 2.3.1. Cassava mealybug infested shoot

(*Epidinocarsis*) *lopezi* and the coccinellids *Diomus* sp. (Neuenschwander, *et al.*, 2003). Of these, introduction of the parasitoid *A. lopezi* has resulted in suppression of this mealybug in Africa, and similar results are expected from releases of this parasitoid in Thailand in 2009 (Winotai *et al.*, 2010).

Other major pests are:
Coconut scale, *Aspidiotus destructor* (See Pests of Papaya)
Spiraling whitefly, *Aleurodicus dispersus* (See Pests of Guava)
Spider mites, *Tetranychus* spp. (See Pests of Eggplant)

2.4 Pests of Cucurbits (Cucurbitaceae)

The family Cucurbitaceae, also known as the gourd family, includes many crops that are predominantly grown in the tropics such as pumpkin, bitter melon, ribbed gourd, snake gourd, watermelon, cucumber, squash, cantaloupe, and others. Cucurbits are often referred to as vine crops. Some are important crops in the drier regions of the tropics as they tolerate high temperatures. The origin of cucurbits varies depending upon the genera. For example, watermelon, cantaloupe, and honeydew melon are from Africa; cucumber, bitter gourd, and snake gourd from India; bottle gourd from Indonesia; and chayote and pumpkin from the New World (George, 2011). Some of the major pests of cucurbits are the melonfly, pumpkin beetles, whiteflies, and aphids.

Common name: Pumpkin beetles
Scientific name: *Aulacophora* spp.
Order: Coleoptera
Family: Chrysomelidae
Distribution:

- *A. femoralis* – Myanmar and Vietnam
- *A. flavomarginata* – Malaysia, Brunei, and Indonesia
- *A. foveicollis* – Africa, Asia, and Europe
- *A. frontalis* – Thailand, Laos, Vietnam, and Singapore
- *A. lewisii* – Vietnam and Malaysia
- *A. similis* – South and Southeast Asia
- *A. coffeae* – Malaysia

Hosts: Cucurbitaceous crops
Life cycle: Adults are known to live 100–200 days. Eggs are yellow and laid singly or in masses of two to five in the soil near the base of the plant. The incubation period ranges from 8–15 days. Newly hatched grubs are pale yellow and turn orange-yellow when full-grown. Grubs bore into roots, and the larval stage lasts for 18–35 days. Pupation takes place in the soil within a cell. Pupae are elongate and pale yellow, and the pupal stage ranges from 4–14 days. The duration of life cycles varies among different species and also depends on host plants and the environment (Figs 2.4.1, 2.4.2, 2.4.3).

Fig. 2.4.1. Adult beetle of *Aulacophora similis*

Fig. 2.4.2. Adult beetle of *Aulacophora flavomarginata*

Fig. 2.4.3. Adult beetle of *Aulacophora* sp. on a melon shoot

Damage: Adults cause serious damage by feeding on leaves and flowers. Seedlings may be heavily attacked to the point that they are totally destroyed. Grubs bore into the roots, causing them to swell, be discolored, misshapen, and in some cases, plant growth is retarded or plants are killed.

Natural enemies: Little or no information is available on the natural enemies of this pest.

Common name: Melon fly
Scientific name: *Bactrocera cucurbitae* (Coquillett)
Order: Diptera
Family: Tephritidae
Distribution: Asia and Africa (Waterhouse, 1993a)
Hosts: Cucurbits, tomato, capsicum, beans, passion fruit, and eggplant. Over 125 hosts have been recorded.
Life cycle: Adults require protein for sexual maturity and in order to develop eggs. Mating is necessary for the production of fertile eggs. Each female is capable of laying over 1,000 eggs. The female inserts her ovipositor directly into the host fruit or vegetable and lays her eggs. There are three larval instars and a prepupal stage. Pupation takes place in the soil beneath the host plant (Waterhouse, 1993a). Males respond to cuelure (Fig. 2.4.4).
Damage: The melon fly is the most important fruit fly pest of vegetable crops, especially of cucurbits. In addition to causing direct damage to fruit, its occurrence in a country leads to quarantine restrictions.
Natural enemies: The parasitoids *Fopius arisanus, Diachasmimorpha dausii, Psyttalia (Opius) fletcheri, Dirhinus anthracina, Spalangia endius, Spalangia hirta, Tetrastichus giffardianus,* and *Tetrastichus dacicida* were recorded. Predators recorded were *Chelisoches morio, Zelus renardii, Pheidole megacephala, Solenopsis geminata, Philonthus turbidus* and *Argiope* spp. Of these, only *P. fletcheri* has given moderately satisfactory control.

Fig. 2.4.4. Adult of *Bactrocera cucurbitae*

Common name: Cucumber moth
Scientific name: *Palpita* (*Diaphania) indica* (Saunder)
Order: Lepidoptera
Family: Pyralidae
Distribution: Asia, Africa, Australia, and Pacific
Hosts: Cucurbitaceous plants
Life cycle: Adult moths have white wings with broad black bands along the anterior margin of the forewings and on the outer margin of fore- and hind wings. Eggs are laid singly or in small clusters on the lower surface of leaves. The incubation period for eggs is about 4–5 days. The caterpillars are green with white median stripes and they feed on the underside of leaves. There are five instars, and the larval period lasts 8–19 days. Pupation takes place inside the rolled leaves, and the pupal period is about 6–7 days.
Damage: Caterpillars mostly feed on leaves and sometimes cause defoliation. Rarely do they attack fruits (Fig. 2.4.5).
Natural enemies: The parasitoids, *Apanteles machaeralis, Apanteles taragamae, Elasmus indicus, Dolichogenidea stantoni, Goniozus sensorius, Phanerotoma hindecasisella,* and a nucleopolyhedrovirus (NPV), were reported attacking larvae.

Common name: Melon thrips
Scientific name: *Thrips palmi* Karny
Order: Thysanoptera
Family: Thripidae
Distribution: Asia, Pacific, Caribbean, Central and South America, Florida (Capinera, 2001)

Fig. 2.4.5. Larva of *Palpita indica* on bitter melon fruit

Host plants: Tobacco, cotton, cucurbits, crucifers, eggplant, bean, soybean, peanut, avocado, mango
Life cycle: Adults are pale yellow, females about 1.16 mm in length and the males 0.91 mm. Eggs are laid in young leaves or flower buds, and they hatch in about 4 days. There are two nymphal stages and in these stages they feed actively along the leaf midribs and veins. After the second instar, they undergo prepupal and pupal stages in the soil. The time taken from egg-laying to adult emergence is about 11 days in the tropics.
Damage: Feeding by adults and nymphal stages on plants causes flower bud drop, stunting of terminal shoots, and curled and scarred fruits. The most serious damage caused by this thrips is transmission of tospoviruses. *Peanut bud necrosis virus* on peanuts, tomato, potato, soybean, and sunflower; *Capsicum chlorosis virus* on peppers, *Iris yellow spot virus* on tomato and peanut; *Melon yellow spot virus*; *Calla lily chlorotic spot virus*; *Sunflower necrosis virus*, *Watermelon silver mottle virus*; and *Watermelon bud necrosis virus* are all transmitted by *T. palmi*. It is the predominant thrips vector species in tropical and subtropical Asia (Pappu *et al.*, 2009).
Natural enemies: The predatory bugs, *Orius mexidentex* and *Carayonocoris indicus*, were reported to feed on *T. palmi* in India. Predators recorded in Hawaii were *Franklinothrips vespiformis*, minute pirate bug, *Orius insidiosus*, *Curinus coeruleus*, *Rhinacoa forticornis*, *Paratriphleps laevisculus*, and *Ceranisus menes*. The predatory bug, *O. mexidentex*, was reported from Thailand (Kumar and Ananthakrishnan, 1984). *Orius strigicollis* and *Haplothrips brevitubus* were recorded from Japan (Kakimoto *et al.*, 2008). The fungi, *Beauveria bassiana*, *Neozygites parvispora*, *Verticillium lecanii*, and *Hirsutella* sp. are known to attack *Thrips tabaci* (Capinera, 2001) and may also infect *T. palmi*.

Common name: Leaf-footed bug
Scientific name: *Leptoglossus gonagra* (F.)
Order: Hemiptera
Family: Coreidae
Distribution: Southeast Asia, Pacific Islands
Hosts: Polyphagous, mostly on cucurbits and morning glory plants.
Life cycle: The life cycle is completed in about 40 days on fruit crops and cucurbitaceae. This pest is commonly found on bitter gourd (*Momordica charantia*). Eggs are laid in strings of varying numbers. The egg stage lasts 4–8 days. There are five nymphal stages that last about 30 days. Adults are 17–20 mm long (Fig. 2.4.6).
Damage: High populations have been observed in cucumbers in Indonesia. Feeding on the young fruit causes distortion as fruit matures.
Natural enemies: Little is known about natural enemies of this pest.

Other pests of regional importance are the stem borer, *Melittia eurytion* (Westwood) (Lepidoptera: Sesiidae), the snake gourd semilooper, *Anadevidia (Plusia) peponis* (Fabricius) (Lep.: Noctuidae), and the bottle gourd plume moth, *Sphenarches caffer* Zeller (Lep.: Ptreophoridae)

Fig. 2.4.6. Leaf-footed bug Leptoglossus gonagra

2.5 Pests of Eggplant (Brinjal) (*Solanum melongena* L., Solanaceae)

India is the center of origin of eggplant. In 2009, China produced around 20 m tonnes, and India 10 m tonnes out of the total world production of about 35 m tonnes, grown over 2 m hectares. This is one of the top ten vegetable crops grown in the world. It is rich in niacin, folic acid, and potassium. It grows well under high rainfall as well as in dry conditions with irrigation and high temperatures. Eggplant fruit and shoot borer is the major constraint in eggplant production in Asia. Misuse of pesticides is a common phenomenon in eggplant cultivation in Asia, and farmers in some countries apply them more than 100 times in a cropping season. Other serious pests are leafhoppers and mites.

Common names: Eggplant fruit and shoot borer, Brinjal fruit and shoot borer
Scientific name: *Leucinodes orbonalis* Guenée
Order: Lepidoptera
Family: Pyralidae
Distribution: South and Southeast Asia, Africa, and Saudi Arabia. Absent in Australia, Pacific Islands, Europe, Americas
Hosts: Eggplant, potato (shoots only), and other *Solanum* spp.
Life cycle: Adult females lay creamy white, flat, oval eggs either singly or in batches of 2–4 on tender shoots, flower buds, and calyces of young fruit. Each female lays about 200 eggs. The egg stage lasts 3–5 days. Upon hatching, the larvae bore into the shoots, flower buds, or fruits, the latter being preferred. Young larvae are creamy white and the mature ones are pinkish. There are five larval instars which are completed in 11–14 days. Full-grown larvae are about 18 mm and emerge from the damaged fruit to pupate within a tough silken cocoon on the fruit, stem, leaf litter or in the soil. The pupal stage lasts 6–8 days (Figs 2.5.1, 2.5.2, 2.5.3)
Damage: Feeding within the fruit causes the most serious damage while boring in the shoot causes wilting and eventual death of the shoot. Damaged buds are shed and damaged fruit is unmarketable (Figs 2.5.4, 2.5.5)

Fig. 2.5.1. Adult moth of *Leucinodes orbonalis*

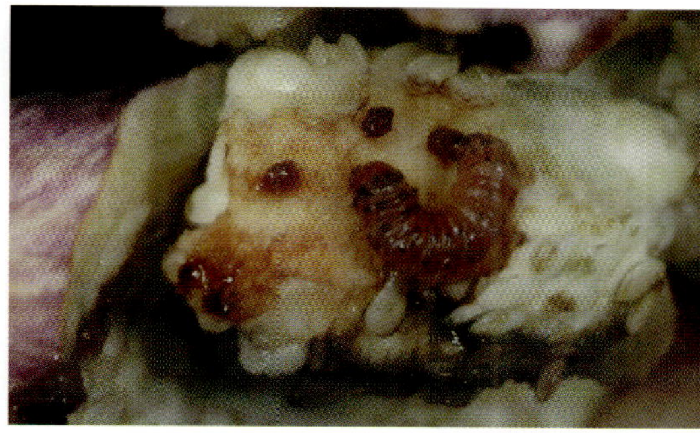

Fig. 2.5.2. *Leucinodes orbonalis* larva feeding in the eggplant fruit

Fig. 2.5.3. *Leucinodes orbonalis* larva in the eggplant stem

Fig. 2.5.4. *Leucinodes orbonalis* damaged fruit of eggplant

Fig. 2.5.5. Wilting of the eggplant shoot caused by *Leucinodes orbonalis*

Natural enemies: Parasitoids recorded are: *Amobia* sp., *Pachyophthalmus* sp., *Cremastus hapaliae*, *Trathala flavoorbitalis*, and *Trathala striata* in Malaysia; *Apanteles* sp., *Chelonus* sp., *Brachymeria lasus*, *T. flavoorbitalis*, and *Xanthopimpla punctata* in the Philippines. An earwig larval predator, *Euborellia annulata*, is being evaluated as a possible biological control agent of this pest in the Philippines. Parasitoids reported from India include *Pseudoperichaeta* sp., *Sturmia parachysops*, *Bracon greeni*, *Chelonus* sp., *Phanerotoma* sp. nr. *hindecasisella*, *Eriborus argenteopilosus*, *Eriborus sinicus*, *Itamoplex* sp., *Pristomerus testaceus*, *T. flavoorbitalis* and *Diadegma apostata*. Two baculoviruses have been reported infecting larvae (Waterhouse, 1998; Srinivasan, 2008).

Leucinodes orbonalis is closely related to the Central American pest, *Neoleucinodes elegantalis*, which is widespread in that region on eggplant (Waterhouse, 1993b, 1998). Parasitoids of *N. elegantalis* may be considered for *L. orbonalis* control and vice versa. Natural enemies of *N. elegantalis* reported are *Copidosoma* sp., *Lixophaga* sp., *Calliephialtes* sp., *Trichogramma* sp., and a fungus, *Beauveria* sp.

Other remarks: Adoption of resistant or moderately resistant varieties such as EG058, Pusa Purple Long, Pusa Purple Cluster, Pusa Purple Round, H-128, H-129, Aushey, Thorn Pendy, Black Pendy, H-165, H-407, Dorley, PPC-17-4, PVR-195, Shyamla Dhepa, Banaras Long Purple, Arka Kesav, Arka Kusmakar, Punjab Barsati, Punjab Chamkila, Kalyanpur-2, Gote-2, BARI Begun-6, BARI Begun-7, and BARI Begun-8 may help in reducing damage by this pest. Removal and destruction of wilted shoots have been shown to be effective in reducing populations of this pest, especially when practiced on an areawide basis (Parker *et al.*, 1995; Srinivasan, 2009). Weekly releases of the egg parasitoid, *Trichogramma chilonis*, at one gm. of parasitized eggs/ha/week, and the larval parasitoid *Bracon hebetor*, at 800–1000 adults/ha/week, are recommended in Bangladesh (Alam *et al.*, 2006).

Common name: Leaf-eating ladybird beetles
Scientific name: *Epilachna vigintioctopunctata* (F.)
Order: Coleoptera
Family: Coccinellidae
Distribution: South and Southeast Asia
Hosts: Eggplant, tomato, potato
Life cycle: Spindle-shaped eggs are laid upright in clusters of 15–50 each, usually on the undersurface of the leaves. The egg stage lasts 2–4 days, and development of four larval instars is completed in 16–18 days. Pupation occurs on the plant, and adults emerge about 4–6 days later (Figs 2.5.6, 2.5.7, 2.5.8, 2.5.9).
Damage: Adults and larvae feed on leaves of plants belonging to the families Cucurbitaceae and Solanaceae. Adults feed both on the upper and lower surfaces of leaves, but larvae are confined to the lower surface. Larvae have a characteristic feeding pattern on the surface of the leaves, scraping away the green tissue and leaving behind windows with leaf veins showing. Leaves that are skeletonized usually dry out.
Natural enemies: Besides heavy rains that reduce field populations of the beetle, natural enemies are important in keeping this insect in check. The egg parasitoid *Tetrastichus* sp. has been reported to kill 70% of the eggs while 30–55% of larvae and

Fig. 2.5.6. Adults of *Epilachna* sp.

Fig. 2.5.7. Eggs of *Epilachna*

Fig. 2.5.8. *Epilachna* grubs feeding on an eggplant leaf

Fig. 2.5.9. *Epilachna* pupae

pupae are attacked by another eulophid, *Pediobius* sp., in Indonesia. *Pediobius foveolatus* is known to occur in the Philippines and India. Several other parasitoids, predators, and pathogens are known to attack this beetle (Esguerra and Gabriel, 1969).

Other remarks: This beetle is not an important pest, but occasionally high populations may develop, which can cause significant defoliation. Some authors have separated the species into *E. vigintioctopunctata* which feeds on solanaceous plants and *E. cucurbitae* which feeds on cucurbitaceous plants (Richards, 1983).

Common name: Melon aphid or cotton aphid
Scientific name: *Aphis gossypii* Glover
Order: Hemiptera
Family: Aphididae
Distribution: Considered cosmopolitan, except for parts of Canada and Asia

Hosts: This aphid attacks a wide variety of plants and is considered the major aphid pest on many crops.

Life cycle: Both winged and wingless forms occur. In tropical areas, female aphids give birth to live nymphs, producing up to 22 per day. There are four nymphal stages with each stage lasting from 1 to 3 days, for a total nymphal period of about 12 days. Adults are light to dark green and nymphs are usually yellow (Figs 2.5.10, 2.5.11).

Damage: This pest sucks sap from its hosts, usually from the undersides of leaves, which causes leaves to cup downwards and appear wrinkled. Young buds and upper leaf surfaces are covered when populations are dense. When young plants are attacked, they may have reduced or stunted growth. The most serious damage by this pest is caused by a wide range of plant viruses that it vectors. In addition, this aphid can produce copious amounts of honeydew that induces sooty mold which can reduce plant photosysthesis and/or diminish fruit quality.

Natural enemies: There are several generalist predators that attack this pest, including coccinellids, syrphids, chrysopids, and hemerobiids. Parasitoids that have been reported from this aphid are: *Lipolexis scutellaris* in Malaysia and the Philippines; *Trioxys communis* in the Philippines; *Aphidencyrtus* sp. in Malaysia; *Lysiphlebia mirzai* in Vietnam; *Aphelinus abdominalis, Aphelinus gossypii, Aphelinus mali, Aphidius colemani, Aphidius gifuensis, Aphidius matricariae, Aphidius similis, Aphidius uzbekistanicus, Cristicaudus nepalensis, Diaeretiella rapae, Ephedrus persicae, Ephedrus plagiator, Lipolexis gracilis, L. scutellaris, Lysaphidus schimitscheki, Praon abjectum, Praon absinthii, Praon myzophagum, Toxares macrosiphophagum, Trioxys acalephae, Trioxys basicurvus, Trioxys equates, Trioxys indicus, Trioxys rietscheli, Trioxys rubicola* in India; and *Aphidius colemani, A. matricariae, Lysiphlebus fabarum, Trioxys sinensis* in Pakistan.

Fig. 2.5.10. *Aphis gossypii* adults, nymphs, and exuviae

Fig. 2.5.11. *Aphis gossypii* on okra fruit

Common names: Brinjal lacewing, eggplant lacewing
Scientific name: *Urentius hystricellus* Richter (*Urentius sentis* Distant)
Order: Hemiptera
Family: Tingidae
Distribution: India, Pakistan, Bangladesh, Sri Lanka, Malaysia, Saudi Arabia, Singapore, Thailand, Yemen, Egypt, Ethiopia, Ghana, Kenya, Niger, Nigeria, Senegal, South Africa, Sudan, Tanzania, Uganda, Zimbabwe
Hosts: Eggplant
Life cycle: This is a small brown lace-bug. Its body is covered with spines, and the wings show a distinct lacelike appearance. Adults live for 30–40 days and lay 35–44 eggs singly in the tissues on the undersurface of the leaves. Eggs hatch in 2–12 days. Nymphs resemble adults, but are wingless. There are five nymphal stages that last 10–23 days. Both adults and nymphs are usually found in groups on the underside of leaves. They suck sap from the leaves, causing whitish to yellowish mottled patches on the leaves.
Damage: In cases of serious infestations, the leaves turn entirely yellow and drop off. Attacked leaves are speckled with small black shiny spots, which are the feces of the bugs.

Common name: Leafhopper
Scientific name: *Amrasca bigutata bigutata* Ishida (= *Amrasca devastans* Distant)
Order: Hemiptera
Family: Cicadellidae
Distribution: South and Southeast Asia
Hosts: Eggplant, okra, cotton, potato
Life cycle: Adults are light green with a black spot on each forewing. Eggs are laid along the midrib and veins of the leaves. Egg duration is about 4–11 days. Nymphs resemble adults but lack wings. Nymphal duration varies from 7–28 days (Fig. 2.5.12).

Fig. 2.5.12. Adult *Amrasca bigutata*

Damage: Both nymphs and adults feed by sucking the sap on the undersurface of leaves. Feeding damage causes yellow spots on the leaves and in severe cases results in crinkling, cupping, bronzing, and a condition known as "hopper burn."
Natural enemies: The parasitoids, *Anagyrus flaveolus* and *Stethynium triclavatum*, have been recorded.

Common name: Spider mites
Scientific name: *Tetranychus* spp. (*Tetranychus urticae* Koch)
Order: Aranae
Family: Tetranychidae
Distribution: Cosmopolitan
Hosts: Polyphagous
Life cycle: Adults vary in color from green to yellow to brown. Eggs are round, white or cream, and hatch in 2-4 days. Upon hatching, they pass through one larval and two nymphal stages before reaching the adult stage. The life cycle is completed in 1–2 weeks.
Damage: Feeding of the mites causes white or yellow specks on the leaves. As the name implies, the mites produce webbing on the leaf surface and, under high population densities, they use strands of web and wind currents to disperse from one plant to another by a process known as "ballooning."
Natural enemies: Predatory mites such as *Phytoseiulus persimilis*, *Metaseiulus occidentalis*, *Amblyseius* spp., and lacewing, *Chrysoperla carnea*, are known to feed on spider mites.

Common name: Leaf roller
Scientific name: *Eublemma olivacea* Walker
Order: Lepidoptera
Family: Noctuidae
Distribution: South Asia
Life cycle: The adult is a medium-sized olive green moth, and it lays eggs in groups of 10–20 on younger leaves. The egg stage lasts for 3–5 days. The larva is purple and feeds on leaves by remaining inside the longitudinally-folded leaves. Larval duration is about 18 days. Pupation takes place inside the folded leaves. Pupal duration is about 7–10 days.
Damage: A minor pest, but larvae do feed on the plant leaves.

Common name: Stem borer
Scientific name: *Euzophera perticella* Ragonot
Order: Lepidoptera
Family: Pyralidae
Distribution: South Asia
Hosts: Eggplant, tomato, potato, peppers
Life cycle: The moth is medium-sized and lays single eggs on young leaves. Eggs are flat and hatch in 3–10 days. Larvae are white with bristly hairs, and the larval duration is about 1–2 months. Larvae pupate in a silken cocoon in the tunnel and adults emerge in 1–2 weeks.

Damage: Tunneling of the larva in the stem causes stunted growth or death of the plant.
Natural enemies: None reported.

Common name: Eggplant budworm
Scientific name: *Scrobipalpa blapsigona* Meyrick
Order: Lepidoptera
Family: Gelechidae
Distribution: South Asia, Africa
Life cycle: The adult is a small brown moth with fringed wings, and the larva is white with a pink tinge.
Damage: Shriveling and dropping of buds
Natural enemies: None reported.

Other major pests are:
Melon thrips, *Thrips palmi* (See Pests of Cucurbits)
Sweet potato whitefly, *Bemisia tabaci* (See Pests of Okra)
Garden looper *Chrysodeixis chalcites* (See Pests of Beans)

Other pests reported from different regions are: ash weevils, *Myllocerus subfasciatus*, *Myllocerus discolor*, *Myllocerus viridanus*, *Myllocerus maculosus* (Coleoptera: Curculionidae); brown leafhopper, *Hishimonus phycitis* (Distant) (Hemiptera: Cicadellidae); blister beetle, *Mylabris pustulata* Thunberg (Coleoptera: Meloidae).

2.6 Pests of Okra (*Abelmoschus esculentus* L., Malvaceae)

Okra is of Ethiopian origin, but it is widely cultivated throughout the tropical and subtropical parts of the world. It is grown mostly for its semimature pod. In some parts of the world, its leaves and dried pods are used in cooking. It is one of the more nutritious vegetables grown. Major pests of okra are the okra fruit and shoot borer, sweet potato whitefly and its biotypes, and leafhoppers.

Common name: Okra fruit and shoot borer
Scientific names: *Earias vittella* (F.) (= *Earias fabia*), *Earias insulana* Boisduval
Order: Lepidoptera
Family: Noctuidae
Distribution: Asia, Australia, Pacific Islands, and Africa
Hosts: Okra and other members of the family Malvaceae
Life cycle: Eggs are laid singly on shoots, flower buds, and young fruit. Eggs hatch in about 3 days, and the larvae are dirty brown with white spots. Larval duration is about 20 days, and pupation occurs in a leathery brown cocoon attached to the shoots, fruits, or leaves. The moth emerges from the pupal case in about 10 days (Figs 2.6.1, 2.6.2).
Damage: This borer is one of the most serious pests of okra, as a single larval bore hole renders the fruit unmarketable. Also, larval boring into stems kills shoot tops and retards plant growth.

Fig. 2.6.1. Adult of *Earias vittella*

Fig. 2.6.2. Larva of *Earias vittella*

Natural enemies: In India, the egg parasitoid, *Trichogramma chilonis* Ischii, larval parasitoids *Bracon lefroyi, Bracon greeni, Bracon hebetor, Bracon brevicornis, Rogas aligarhensis, Rogas testaceus, Bassus* sp., *Apanteles* sp., *Phanerotoma hindecasisella, Elasmus* sp., *Elasmus johnstoni, Actia aegyptia, Actia hyalinata,* and *Polyodaspis* sp., and pupal parasitoids, *Chelonus* sp., *Chelonus rufus, Brachymeria lasus, Brachymeria nephantidis, Chalcis tachardiae, Chalcis* sp., *Centrochalcis* sp., and *Goryphus nursei* have been reported. The wasp, *Eumenes petiolata,* was found stocking its nests with the caterpillars of *E. vittella*. In Malaysia, a braconid parasitoid was recorded. Additionally, a predatory lacewing has been reported to feed on eggs of *E. vittella*.

Common names: Sweet potato whitefly, cotton whitefly, tobacco whitefly (*B. argentifolii* or Biotype B – silverleaf whitefly)
Scientific name: *Bemisia tabaci* (Gennadius)
Order: Hemiptera
Family: Aleyrodidae
Distribution: Tropical and subtropical areas of the world
Hosts: These whiteflies have a wide host range of over 500 herbaceous annuals including cotton, okra, tomato, eggplant, peppers, beans, cassava, cucurbits, potato, sweet potato, and crucifers (De Barro, 1995; Bosco and Caciagli, 1998).
Life cycle: Pear-shaped eggs are laid in an upright position on the undersurface of the leaf. Each female lays about 200 eggs, and the incubation period is about 7 days. Upon emergence, nymphs move a short distance and settle down for feeding after which they do not move. Nymphs are flat, scalelike, somewhat spiny, and covered with wax. There are four instars and the fourth one is known as the pupa. It takes about 2–4 weeks for nymphal development.
Damage: This insect is considered to be one of the world's most destructive pests. Since the early 1980s, it has become a major pest of vegetables. It causes physical damage by sucking the plant sap and coating the plant with honeydew, resulting in the growth of sooty mold. However, the most damage is caused by vectoring of plant viruses such as *Tomato yellow leaf curl virus* (TYLCV), *Cucumber vein yellowing virus* (CVYV), *Tobacco leaf curl virus* (TLCV), *Okra yellow leaf curl virus* (OYLCV), *Cotton leaf curl virus* (CLCV), *Watermelon chlorotic stunt virus* (WCSV), *Sweet potato chlorotic stunt*

virus (SPCSV), *Cassava mosaic virus* (CaMV), *Watermelon chlorotic stunt virus* (WCSV), *Lettuce infectious yellow virus* (LIYV), *Cucurbit yellow stunting disorder virus* (CYSDV), *Eggplant yellow stunting disorder virus* (EYSDV), *Cucumber bottle gourd mosaic virus* (CBGMV), *Cowpea mild mottle carlavirus* (CMMV), *Sweet potato leaf curl badnavirus* (SPLCV), and others.
Natural enemies: There are 19 described parasitoids belonging to *Encarsia* and *Eretmocerus* spp., as well as many more as yet undescribed that attack the sweet potato whitefly. Commercial preparations of the fungi *Verticillium lecanii, Paecilomyces fumosoroseus,* and *Beauveria bassiana* have proven effective in controlling this pest (Gerling *et al.*, 2001).

Common names: Cotton stainer, red cotton bug
Scientific name: *Dysdercus cingulatus* (F.)
Order: Hemiptera
Family: Pyrrhocoridae
Distribution: India, Bangladesh, China, Japan, Southeast Asia, Australia, Saipan, Palau, Yap, Solomon Islands, Papua New Guinea, Vanuatu, and New Caledonia (Waterhouse, 1998).
Hosts: Cotton, okra, kapok, rosella, hibiscus, hemp, and other plants of the families Malvaceae and Bombacaceae
Life cycle: Eggs are laid in clusters of about 80–100 in the soil under host plants, in small superficial holes. High humidity is required for egg development. The egg stage lasts 6 days. The five nymphal stages are completed in 25–27 days. Adults are 11–17 mm in length. Adults and nymphs produce defensive secretions (Figs 2.6.3, 2.6.4).
Damage: Older nymphs and adults suck sap from pods and seeds.
Pest status: This bug can be a serious pest of okra under certain conditions.
Natural enemies: The pyrrhocorid predator, *Antilochus coquebertii*, has been reported attacking this pest in India, the Philippines, and Malaysia. Spiders, carabid beetles, and reduviid bugs also prey on this bug.

Fig. 2.6.3. *Dysdercus cingulatus* adult

Fig. 2.6.4. *Dysdercus cingulatus* nymphs

Common name: Cotton leaf roller
Scientific name: *Sylepta derogata* (F.)
Order: Lepidoptera
Family: Pyralidae
Distribution: Asia, Africa, Australia
Hosts: Cotton, durian, okra, and long bean
Life cycle: Flat yellow eggs are laid singly on the lower surface of leaves. Eggs hatch in about 3 days. Young larvae feed on the lower surfaces of the leaves and the older ones roll the leaves and feed from within. Pupation takes place within the leaf roll. The pupal period is 6–9 days and the total life cycle averages 25 days.
Damage: This insect is a sporadic pest of malvaceous plants. Severe attack results in occurrence of numerous leaf rolls and defoliation. Under normal conditions, natural enemies keep this pest under control.
Natural enemies: Larval parasitoids reported from the Philippines include *Brachymeria obscurata, Elasmus philippinensis, Pediobius* sp., and *Xanthopimpla punctata*; and from India, *Brachymeria bengalensis pulchellae, Elasmus indicus, Enicospilus atricornis, Goryphus nursei, Microtoridea lissonota, X. punctata, Neopimploides syleptae, Apanteles* sp., *Bracon lefroyi, Microbracon recinicola, Phanerotoma hindecasisella, Trichospilus pupivora*, and *Bessa remota*. Eggs are parasitized by *Trichogramma* sp. and pupae by *Xanthopimpla* sp. in India.

Common name: Cotton semilooper
Scientific name: *Anomis flava* (F.) (= *Cosmophila flava*)
Order: Lepidoptera
Family: Noctuidae
Distribution: Asia, Africa
Life cycle: Eggs are yellow, rounded on the top and flat underneath. They are laid singly on the lower surface of leaves and sometimes along the stem and young shoots of okra. The incubation period is 2–3 days. The larva is a semilooper and when full-grown it measures 33–40 mm in length. There are five larval stages that are completed in 12–13 days. Pupation takes place in a leaf fold held together by silk. The pupal stage lasts about 7 days (Figs 2.6.5, 2.6.6).

Fig. 2.6.5. Adult male of *Anomis flava*

Fig. 2.6.6. *Anomis flava* larva feeding on an okra leaf

Damage: Larvae are foliage feeders and are capable of total defoliation during severe infestations. When leaves are not available, larvae may feed on young shoots, flower buds, and fruit. When young seedlings are defoliated, they may not recover.
Natural enemies: An egg parasitoid, *Trichogramma minutum*, larval parasitoids, *Euplectrus manilae* and *Apanteles* sp., and a pupal parasitoid, *Brachymeria lasus* (= *Brachymeria obscurata*) have been recorded on this pest in the Philippines. Several other parasitoids and predators were recorded in other parts of the world.

Common name: Pink hibiscus mealybug
Scientific name: *Maconellicoccus hirsutus* (Green)
Order: Hemiptera
Family: Pseudococcidae
Origin: Southeast Asia
Distribution: South and Southeast Asia, Caribbean, Florida, California, Middle East
Host: Hibiscus, okra, coffee, citrus, guava, beans, cotton, cocoa, others
Life cycle: The adult is pink with a white waxy coating. Eggs are laid in an egg sac covered with wax. Each female lays about 600 eggs. First instar nymphs (crawlers) are dispersed by walking and wind. This mealybug completes about 15 generations a year in the tropics (Fig. 2.6.7).
Damage: This pest feeds on tender shoots by sucking the sap and the toxic saliva injected into the tissue causes curling and distortion. Infested buds may not open and fruits can be deformed. This mealybug also produces copious honeydew which supports growth of black sooty mold.
Natural enemies: The parasitoids *Anagyrus kamali* Moursi, *Gyranusoidea indica*, and *Allotropa* sp. near *mecrida* and the predators *Cryptolaemus montrouzieri* and *Scymnus coccivora*, have been used for classical biological control of this pest in various countries (Reddy *et al.*, 2009).

Fig. 2.6.7. Pink hibiscus mealybug infested okra shoot

Other major pests are:
Leafhopper, *Amrasca bigutata bigutata* (See Pests of Eggplant)
Cotton bollworm, *Helicoverpa armigera* (See Pests of Tomato)
Cotton aphid, *Aphis gossypii* (See Pests of Eggplant)

Other pests of regional importance are the shoot weevil, *Alcidodes affaber* Auriv (Col.: Curculionidae), the stem weevil, *Pempherulus affinis* Fst. (Col.: Curculionidae), the dusky cotton bug, *Oxycarenus laeus* Kirby (Hem.: Lygaeidae), and the stem fly, *Melanagromyza obtusa* (Mall.) (Dip.: Agromyzidae).

2.7 Pests of Onion (*Allium cepa* L., Alliaceae)

Onion ranks second only to tomato among cultivated vegetables. Onion originated in the region of Afghanistan, Iran, and Pakistan (George, 2011). Main onion producing countries in Asia are China, India, Japan, Pakistan, and South Korea. Several types of onions are grown all over the world and some are used both green and dry. Onions have become popular because of their medicinal value. Onion thrips that transmit virus diseases in addition to direct damage by army worms, onion fly, and leaf miners are the most serious insect pests of onions in tropical Asia.

Common name: Onion thrips
Scientific name: *Thrips tabaci* Lindeman
Order: Thysanoptera
Family: Thripidae
Origin: Middle East, countries bordering the Black and Caspian Seas
Distribution: Cosmopolitan
Hosts: More than 300 species of plants have been recorded for this pest. Onion, in particular, and other closely-related species are preferred.
Life cycle: Adults are yellow-brown with dark transverse bands across the thorax and abdomen. They live 1–4 months depending on the temperature and humidity. The white eggs are inserted singly in the epidermis of the leaves and hatch in 4–10 days. Two nymphal instars and adults feed on the plant by rasping and sucking and the two pupal instars are found in the soil. The life cycle is completed in 3–5 weeks depending on temperature (Lewis, 1997) (Fig. 2.7.1).
Damage: Feeding of *T. tabaci* causes silvery sheen on the leaves due to draining of the cell contents, resulting in air-filled spaces. More important than the direct damage is the transmission of tospoviruses such as *Tomato spotted wilt virus*. This virus causes serious economic loss to several vegetable crops throughout the world. Larval stages of *T. tabaci* obtain and carry the tomato spotted wilt virus but transmission occurs only after thrips reach adulthood. Other tospovirus diseases transmitted by *T. tabaci* include *Iris yellow spot virus* (IYSV) in onions, *Calla lily chlorotic spot virus*, *Melon yellow spot virus*, *Peanut bud necrosis virus*, and *Tomato yellow fruit ring virus* (Pappu *et al.*, 2009).
Natural enemies: The eulophid parasitoids *Ceranisus vinctus* from the Philippines and *Goetheana* (*Dasyscapus*) *parvipennis* from Java have been reported from *Thrips*

Fig. 2.7.1. *Thrips tabaci* on an onion leaf

tabaci. Goetheana parvipennis from Ghana was introduced and established in the West Indies for control of *T. tabaci* and *Selenothrips rubrocinctus*. The phytoseiid mites, *Amblyseius mckenziei*, *Amblyseius cucumeris*, and *Typhlodromips sessor* and the fungal pathogens *Neozygites parvispora* and *Verticillium lecanii* are used in glasshouses for control of *T. tabaci* in Europe.

Common names: Beet army worm, onion army worm
Scientific name: *Spodoptera exigua* (Hübner)
Order: Lepidoptera
Family: Noctuidae
Distribution: Europe, Asia, United States
Hosts: Onion, garlic, leek, cruciferous crops
Life cycle: Clusters of eggs are laid on leaves and covered with white feltlike scales. Eggs hatch in about 3 days. Larvae feed clustered together after hatching, but soon disperse. There are usually six larval instars which complete development in 9–14 days. Pupation occurs in the soil and the pupal period lasts for about 10 days (Figs 2.7.2, 2.7.3, 2.7.4).
Damage: Damage is caused by larvae feeding on the leaves. In onions and shallots, the larvae enter the leaves as early instars and feed from inside the leaves, causing characteristic "windowpaning," leaving only the outer cuticle of the leaf (Fig. 2.7.5).
Natural enemies: A nucleopolyhedrosis virus has been isolated from *S. exigua* in Indonesia and commercial formulations are available for field use. This has helped to reduce dependence on chemical insecticides. The parasitoids *Diadegma semiclausum*, *Eriborus argenteopilosus*, *Charops brachypterum*, and *Euplectrus* cf. *chapadae* were recorded in Indonesia (Shepard and Barrion, 1998). A fungus, *Nomuraea rileyi*, and a microsporidian (protozoan) were also reported from Indonesia. *Chelonus insularis*,

Fig. 2.7.2. Adult of *Spodoptera exigua*

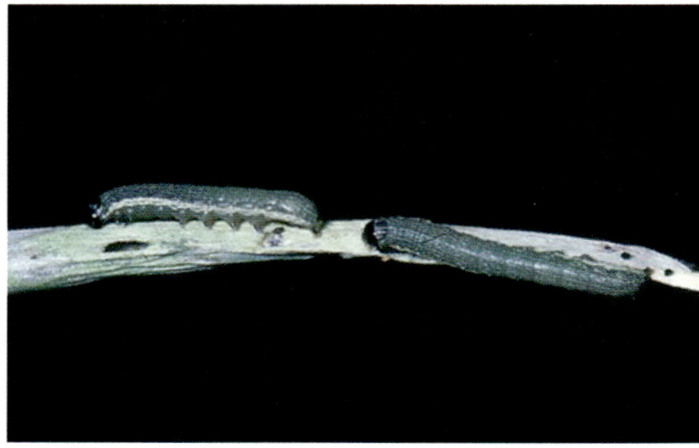

Fig. 2.7.3. Larvae of *Spodoptera exigua* feeding on an onion leaf

Fig. 2.7.4. Egg mass of *Spodoptera exigua*

Fig. 2.7.5. *Spodoptera exigua* damage on onions

Cotesia marginiventris, *Meteorus autographae*, and the tachinid *Lespsia archippivora* have been recorded in the U.S. Predators found attacking the eggs and larvae in the U.S. were minute pirate bugs, *Orius* sp.; big-eyed bugs, *Geocoris* spp.; damsel bugs, *Nabis* spp.; and a predatory shield bug, *Podisus maculiventris* (Figs 2.7.6, 2.7.7, 2.7.8, 2.7.9, 2.7.10).

Fig. 2.7.6. Tachinid parasitoid eggs on *Spodoptera exigua* larva

Fig. 2.7.7. Larva of *Spodoptera exigua* infected by nucleopolyhedro virus

Fig. 2.7.8. Larva of *Spodoptera exigua* killed by nucleopolyhedro virus

Fig. 2.7.9. Larva of *Spodoptera exigua* killed by *Nomurea rileyi*

Fig. 2.7.10. A microsporidian infecting a larva of *Spodoptera exigua*

Other remarks: There are several larval color phases, ranging from green to nearly black. Sometimes the dark-colored larvae are confused with other army worm species. This pest has become resistant to many chemical pesticides.

Common name: Black onion aphid
Scientific name: *Neotoxoptera formosana* (Takahashi)
Order: Hemiptera
Family: Aphididae
Distribution: United States, Mexico, Brazil, Chile, France, Italy, Netherlands, Spain, Japan, China, Taiwan, Korea, Indonesia, Australia, Papua New Guinea, New Zealand
Hosts: Onion, chives, leek, garlic, shallot
Life cycle: This pest completes its nymphal stages in 7 days on shallots (Fig. 2.7.11).
Damage: This aphid forms dense colonies and can reduce the value of the crop due to feeding and because of the physical presence of the aphids, especially on leaf onions.
Natural enemies: No information is available on the natural enemies of this pest but it is likely that generalist predators, such as ladybeetles and lacewing larvae, feed on this aphid.

Other remarks: *N. formosana* is difficult to separate from a closely related black aphid, *N. oliveri* (Essig).

Fig. 2.7.11. The aphid, *Neotoxoptera formosana*, on an onion leaf

Common name: Onion fly
Scientific name: *Delia (Hylemya) antiqua* (Meigen)
Order: Diptera
Family: Anthomyiidae
Distribution: Europe, Cyprus, India, Iran, Israel, Japan, Korea, Turkey, Egypt, Canada, Mexico, North America, Brazil, Colombia
Hosts: Onion, garlic, shallot, leek
Life cycle: White elongate eggs are laid in groups on the shoots, leaves, bulbs, and in the soil at the base of the plants. Eggs hatch in 3–8 days. Maggots keep together and their boring in the stem and bulb causes rotting of the stem at the base and the bulbs. In some cases 50 maggots may be found in one bulb. Larval duration is about 3 weeks and they undergo three molts. Pupation takes place in the soil and the pupal duration is about 3 weeks.
Damage: Distorted growth of bulbs and rotting of the tissue renders them unmarketable. Often the bacterium *Bacillus carotovorus* enters the maggot damaged areas and causes soft rot.
Natural enemies: A nematode, *Heterotylenchus aberrans*, parasitized adult flies in France and Denmark. A parasitoid, *Aphaereta pallipes*, has been reported from Canada.

Other major pests are:
Western flower thrips, *Frankliniella occidentalis* (See Pests of Tomato)
Cluster caterpillar or army worm, *Spodoptera litura* (See Pests of Tomato)

2.8 Pests of Pepper (*Capsicum annuum* L. and *C. frutescens* L., Solanaceae)

Capsicum annuum, commonly known as sweet or bell pepper, is grown as a vegetable. *Capsicum frutescens* is the hot pepper grown as a spice. Both are Meso-American in origin and are cultivated throughout the tropics. The Asian countries India, Indonesia, Korea, China, Bangladesh, Thailand, and Pakistan are the leading producers of pepper in the world. Most important pests are broad mites and the whiteflies that transmits virus diseases.

Common names: Broad mite, Yellow tea mite
Scientific name: *Polyphagotarsonemus latus* (Banks)
Order: Acarina
Family: Tarsonemidae
Distribution: Cosmopolitan
Hosts: Polyphagous. Many vegetable and fruit crops are attacked
Life cycle: Eggs are white, ovoid, and laid singly on the undersurfaces of leaves. Larvae with six legs emerge from the eggs and feed on the leaves until they enter a quiescent or pupal stage. Adults emerge and stay on the lower surface of leaves. The 0.2 mm long females are twice the size of males. Females lay about 8 eggs a day and the life cycle is completed in 4–5 days.
Damage: This mite is a serious pest of peppers, beans, cucurbits, tomato, eggplant, and others. Usually it affects tender leaves in the growing shoots. Affected shoots

Fig. 2.8.1. Pepper plant infested with *Polyphagotarsonemus latus*

become elongated, leaves become curled and distorted, flowers drop, and fruits are malformed and scarified (Fig. 2.8.1).

Natural enemies: Phytoseiid mites and some coccinellids prey on broad mites. In Mauritius, the predatory mite, *Typhlodromus* (= *Amblyseius*) *ovalis*, controls *P. latus* on capsicums. The predatory mite, *Neoseiulus californicus*, has controlled *P. latus* on Chinese eggplant in Honduras.

Common names: Chili thrips, Hevea thrips
Scientific name: *Scirtothrips dorsalis* Hood
Order: Thysanoptera
Family: Thripidae
Distribution: South Africa, South and Southeast Asia, Hawaii, Venezuela, the Caribbean
Hosts: Chili, rubber, cashew, guava
Life cycle: Adults vary from pale to brownish red and eggs are inserted into the plant tissue. The duration of the egg stage is 6–8 days, the larval stage is 6–7 days, and the pupal stage is 2–3 days. Adults live an average of 11 days and feed mostly on leaves.
Damage: Feeding by these thrips causes leaves to curl and drop. Fresh buds also drop when fed upon. Feeding deforms young leaves and stains or scars fruit. Apparently, this pest is capable of spreading *Tomato spotted wilt virus*, *Peanut bud necrosis virus*, *Peanut chlorotic fan virus*, and *Tobacco streak virus*.
Natural enemies: *Franklinothrips megalops* and *Scolothrips indicus* to feed on chili thrips in India (Watson *et al.*, 1995).

Common name: Pepper gallfly
Scientific name: *Asphondylia capsici* Barnes
Order: Diptera
Family: Cecidomyidae
Distribution: Indonesia, India, Cyprus, Turkey
Hosts: Peppers
Life cycle: The life history of this pest is not well known. Eggs are deposited in pepper fruits and larvae develop inside. Often, part of the pupal case can be seen protruding from the pepper fruit.
Damage: Larvae feed inside the fruit of pepper. When young pods are attacked, they become small and deformed. When older fruit is attacked it becomes twisted and deformed (Figs 2.8.2, 2.8.3, 2.8.4).

Fig. 2.8.2. Pepper fruit deformed by *Asphondylia capsici* infestation

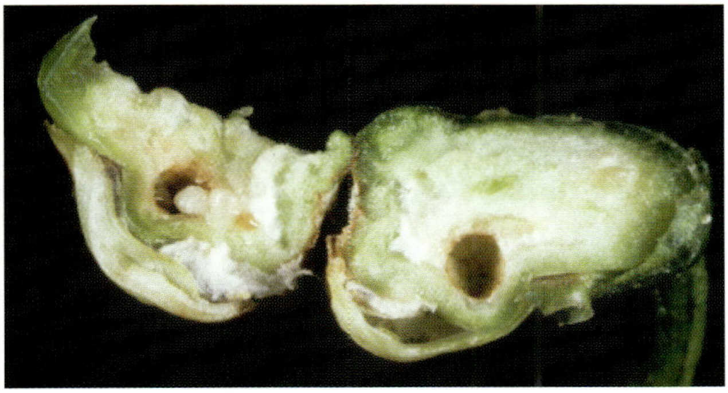

Fig. 2.8.3. *Asphondylia capsici* maggot inside a cut open pepper fruit

Fig. 2.8.4. *Asphondylia capsici* pupal case extruding from the fruit after the adult emergence

Natural enemies: The larval parasitoids *Eurytoma* sp. and *Dinarmus* sp. were reported from Indonesia and India, respectively. The parasitoids *Eurytoma dentata, Pseudocatoloccus nitescens* and *Eupelmus urozonus* were reported from Turkey (Sertkaya *et al.*, 2006).

Other major pests are:
Corn earworm, *Helicoverpa armigera* (See Pests of Tomato)
Spiraling whitefly, *Aleurodicus dispersus* (See Pests of Guava)
Cotton aphid, melon aphid, *Aphis gossypii* (See Pests of Eggplant)
Sweet potato whitefly, *Bemisia tabaci* (See Pests of Okra)

Other pests of regional importance are: Pagoda bagworm, *Pagodiella hekmeyeri* Heylaerts (Lep.: Psychidae), Pistachio red bug, *Spilostethus pandurus* Scopoli (Hem.: Lygaeidae), Green peach aphid, *Myzus persicae* (Sulz.) (Hem.: Aphididae), and *Bactrocera latifrons* (Hendel) (Dip.: Tephritidae)

2.9 Pests of Potato (*Solanum tuberosum* L., Solanaceae)

Potato is the world's fourth largest crop, following rice, wheat, and corn. It is in the nightshade family and was introduced to the outside world from the Andes about four centuries ago. Tubers are often kept in warehouses in cold storage for long periods of time. China is now the major potato-producing country in the world. Other than *Phytophthora infestans*, which caused the famous great Irish famine, the potato tuber moth is the most important pest worldwide.

Common name: Potato tuber moth
Scientific name: *Phthorimaea operculella* (Zeller)
Order: Lepidoptera
Family: Gelechiidae
Distribution: Worldwide
Hosts: Mainly potato, but also attacks tomato and tobacco
Life cycle: Whitish eggs are laid, either singly or in clusters of up to six, directly on the plant or on exposed potatoes. After about a week, the eggs hatch and newly hatched larvae bore into leaves, small stems, or directly into the tuber. Larvae pass through four instars, which takes about 3 weeks. Pupation takes place in the soil or just below the skin of a tuber inside a loosely-formed cocoon. The complete life cycle takes about a month. Moths are short-lived (Figs 2.9.1, 2.9.2).
Damage: Larvae tunnel between the upper and lower epidermis of leaves but then migrate to the tubers where the most serious damage is caused. Feeding on tubers also provides an entry point for pathogens. Besides being a pest in the growing crop, this insect is also an important pest in stored potatoes.
Natural enemies: A granulosis virus has been isolated from larvae. Several parasitoids have been reared from the larvae. These include *Apanteles subandinus, Copidosoma koehleri, Orgilus lepidus, Diadegma pulchripes, Pnigalio pectinicornis* and *Sympiesis viridula*.

Fig. 2.9.1. Potato tuber moth, *Phthorimaea operculella* adult

Fig. 2.9.2. Potato tuber moth, *Phthorimaea operculella* larva and pupa

Common name: Potato aphid
Scientific name: *Macrosiphum euphorbiae* (Thomas)
Order: Hemiptera
Family: Aphididae
Distribution: Worldwide
Hosts: Potato, tomato, eggplant, pepper, pea, bean, apple, sweet potato, and others, including many weeds
Life cycle: These aphids feed and reproduce year-round. No males are known. Wingless females give live birth to about 50 live nymphs which mature in 2–3 weeks. As populations become dense and/or food becomes scarce, winged adults are produced and spread to other areas. After an adequate food source is found, only wingless adults are produced. Many generations are produced per year.
Damage: Damage is caused by removing plant juices with its needlelike sucking mouthparts. Thus, plant vigor is reduced, leaves shrivel, and blooms may fall off, thereby reducing overall yield. It also acts as a vector of several important plant viruses, including various mosaics, leaf roll, spindle tuber, and others. Dense populations cause the build-up of sooty mold which grows on the sugary solution secreted by the aphids.
Natural enemies: These aphids are attacked by a number of generalist predators including ladybeetles, lacewing larvae, syrphid larvae, *Geocoris* spp., and others. A number of parasitoid species also are important natural biological controls. During rainy weather, entomogenous fungi also infect these aphids.

Other major pests are:
Leaf miners: Pea leaf miner, *Liriomyza huidobrensis*; Serpentine leaf miner, *L. sativae*; stone leek leaf miner, *L. chinensis*; Chrysanthemum leaf miner, *L. trifolii* (See Pests of Tomato)
Melon aphid or cotton aphid, *Aphis gossypii* (See Pests of Eggplant)
Green peach aphid, *Myzus persicae* (See Pests of Peppers)
Leaf-eating ladybird, *Epilachna vigintioctopunctata* (See Pests of Eggplant)
Garden looper, *Chrysodiexis chalcites* (See Pests of Beans)
Green vegetable bug or southern green stinkbug, *Nezara viridula* (See Pests of Beans)

2.10 Pests of Sweet Potato (*Ipomoea batatas* (L.) Lam., Convolvulaceae)

Sweet potato originated in South America and is an important food crop of poor farmers of the Old and New World tropics. China produces the most sweet potatoes but India and Indonesia are also major producers in Asia. The sweet potato weevil is a major pest and it is managed by using pheromone traps charged with microbial pesticides and cultural methods in many countries.

Common name: Sweet potato weevil
Scientific name: *Cylas formicarius* (F.)
Order: Coleoptera
Family: Curculionidae
Distribution: Africa, Asia, Australia, United States, Mexico, Panama, Cuba, Jamaica, Puerto Rico, Dominican Republic, Haiti, Guyana, St. Kitts-Nevis
Hosts: Sweet potato, morning glory, *Ipomoea iliate, Ipomoea pes-caprae*
Life cycle: Adults live up to 3 months. Each female lays about 200 eggs at the rate of about 2 eggs per day. Eggs are laid in cavities chewed by the female in the vines or tubers. The egg stage lasts 3–7 days. Larvae tunnel and feed within the vines and tubers. There are three instars and the larval stage lasts from 7 to 11 days. Pupation takes place within the vine or tuber and lasts for 5 to 7 days. The entire life cycle is completed in 19–30 days. The adults can live for a month without food or water and up to 8 months under favorable conditions. They can fly, but rarely do. Dispersal occurs largely when the infested vines or tubers are transported by humans (Figs 2.10.1, 2.10.2).
Damage: Adult weevils feed on leaves, vines, and tubers, and larvae bore into the vines and tubers, making the latter unmarketable. The main damage is caused by larval tunneling in tubers. Even a light infestation is sufficient to render the tubers inedible because of contamination.

Fig. 2.10.1. Sweet potato weevil, *Cylas formicarius*

Fig. 2.10.2. Grub of sweet potato weevil, *Cylas formicarius*

Natural enemies: The parasitoids *Bracon* sp. and *Rhaconotus* sp. from India; *Bracon cylasovorus* from the Philippines; and a fungus, *Isaria* sp., from Java, Indonesia have been recorded from this pest.

Common names: Sweet potato hawkmoth, sweet potato hornworm
Scientific name: *Agrius convolvuli* (Linnaeus)
Order: Lepidoptera
Family: Sphingidae
Distribution: Southern Europe, Africa, Asia, Australia, Pacific Islands
Hosts: *Ipomoea* spp. and other Convolvulaceae; *Vigna* spp., and *Phaseolus* spp.
Life cycle: Eggs are laid singly on leaves and in a few cases on stems. There are five larval instars. Larvae are mostly green, but black and brown forms also occur. Full-grown larvae reach a length of 9 cm and pupation takes place in the soil in an earthen cell. The life cycle is completed in about a month's time (Figs 2.10.3, 2.10.4).
Damage: Caterpillars can defoliate sweet potato and delay harvest. Defoliation of legumes may result in partial crop failure. However, this insect is rarely a serious pest.
Natural enemies: The tachinid flies, *Zygobothria atrcpivora* from Malaysia and *Zygobothria ciliata* from Indonesia and the Philippines, *Trichogramma achaeae, Trichogramma agriae, Trichogramma australicum, Trichogramma confusum* and *Telenomus* sp. from India, *Trichogramma minutum* from Indonesia, *Apanteles* sp. and *Charops bicolor* from China, *Hadrojoppa cognatoria* from Japan, and *Trichogramma chilonis* from Guam have been recorded on *A. convolvuli* (Waterhouse, 1998). A reduviid, *Sycanus* sp., attacks the larvae in Indonesia.

Fig. 2.10.3. Larva of *Agrius convolvuli*

Fig. 2.10.4. Pupa of *Agrius convolvuli*

Common name: Tortoise beetles
Scientific name: *Aspidomorpha miliaris* (F.) and other species listed below
Order: Coleoptera
Family: Chrysomelidae
Origin: South and Southeast Asia
Distribution: Asia
A. miliaris – India, Bangladesh, Indonesia, Malaysia, China, Thailand, Vietnam, Brunei, and the Philippines
Aspidomorpha sanctaecrucis – Indonesia
Aspidomorpha furcata – Cambodia, Vietnam, and Singapore
Cassida (Metriona) circumdata Herbst – the Philippines, India, Indonesia, Nepal, Malaysia, Thailand, Vietnam, Japan, China, and Taiwan
Laccoptera tredecimguttata Wagener – the Philippines
Laccoptera tredecimpunctata Fabricius – Indonesia, Thailand, the Philippines, and Cambodia
Life cycle: *(A. miliaris)*: Adult beetles are pale lemon with black spots on the elytra and are 9–14 mm in length. The female beetle lays spindle shaped eggs in a mass enclosed in an ootheca on the lower surface of the leaves. A female will lay 18–32 egg masses. Eggs hatch in 9–10 days. Young larvae cluster together, feeding on the undersurface of older leaves. Larger larvae disperse over the plant. Larvae are ochraceous with all the exuviae hooked to the sub-anal spines. Larvae undergo five instars and the larval period ranges from 18–22 days. Pupation takes place on the leaf and the pupal duration ranges from 5–7 days. The complete life cycle varies from 4 to 6 weeks (Figs 2.10.5, 2.10.6).
Damage: Feeding by larvae causes round holes in the leaves, but these beetles are not considered major pests.
Natural enemies: Egg parasitoids appear to be common and probably help to keep the population low. In addition, a larval parasitoid is known from Indonesia and India.

Fig. 2.10.5. Tortoise beetle, *Aspidomorpha* sp.

Fig. 2.10.6. Tortoise beetle

Common name: Sweet potato stem borer
Scientific name: *Omphisa anastomosalis* Guenée
Order: Lepidoptera
Family: Pyralidae
Distribution: South and Southeast Asia, Japan, China, Hawaii
Hosts: Sweet potato, *Merremia vitifolia*
Life cycle: The moth has a reddish brown head, thorax, and abdomen, and wings are white with reddish brown markings. Eggs are laid singly in most cases but sometimes two or three are laid on the undersides of leaves and petioles. When eggs hatch, larvae bore into the stems, moving toward the base and, in some cases, reaching the tubers. Pupation takes place in the stem. The life cycle takes about seven to eight weeks (O'Connor, 1969).
Damage: Tunneling of the larvae in stems causes them to wilt and die. Additional damage is done when the larvae tunnel into tubers.
Natural enemies: None recorded

Other major pests are:
Cluster caterpillar or army worm, *Spodoptera litura* (See Pests of Tomato).
Sweet potato whitefly, *Bemisia tabaci* (See Pests of Okra)

Other pests of regional importance are Sweet potato leaf roller, *Brachmia* sp. (Lep.: Gelechiidae), red-striped bug *Physomerus grossipes* (Fabricius) (Hem.: Coreidae) (Fig. 2.10.7), and *Anoplocnemis phasiana* Fabricius (Hem.: Coreidae) (Fig. 2.10.8).

Fig. 2.10.7. *Physomerus grossipes*

Fig. 2.10.8. *Anoplocnemis phasiana*

2.11 Pests of Tomato (*Solanum lycopersicum* (L.), Solanaceae)

Common names: Corn earworm, cotton bollworm, tomato fruitworm
Scientific name: *Helicoverpa armigera* Hübner
Order: Lepidoptera
Family: Noctuidae
Distribution: Europe, Africa, Australia, Asia
Hosts: Tomato, cotton, corn, peppers, eggplant, melons, crucifers, beans, others
Life cycle: Eggs are about 0.5 mm in diameter, spherical in shape, ribbed, and with a flattened surface for attachment. They are laid singly on the upper parts of the plant; eggs are white when first laid, but change through greenish yellow to dark brown before hatching. Eggs hatch in about 3 days and young larvae feed on tender leaves, flowers and terminal shoots. Larvae pass through six instars and the larval period ranges from 17–25 days. Mature larvae vary in color and generally prefer to feed by boring directly into the fruits. Often only the front portion of the body is inside the hole. Larvae are cannibalistic and pupate in soil. The pupal stage lasts for 12–14 days (Figs 2.11.1, 2.11.2).
Damage: This is one of the major pests of food and fiber crops in the world. Damage is mostly due to larval feeding on flowers, flower buds, and fruits.
Natural enemies: Several egg, larval, and pupal parasitoids have been reported from Africa and Asia (Waterhouse and Norris, 1987). Some of the parasitoids recorded from Asia are egg parasitoids, *Trichogramma chilonis* and *Trichogrammatoidea armigera*; larval parasitoids include *Banchopsis ruficornis*, *Campoletis chlorideae*, *Enicospilus* sp., *Eriborus* sp., *Apanteles ruficrus*, and *Bracon brevicornis*, *Carcelia illota*, *Exorista xanthaspis*, *Palexorista imberbis*, and *Winthemia neowinthemoides*. There is also a larval pupal parasitoid *Goniophthalmus halli*, and pupal parasitoid, *Eriborus argenteopilosus*. A fungal disease, *Nomuraea rileyi*, on larva was reported from Indonesia (Fig. 2.11.3). Formulations of nucleopolyhedrovirus of *H. armigera* have been effective in controlling this pest.

Fig. 2.11.1. Adult of *Helicoverpa armigera*

Fig. 2.11.2. Larva of *Helicoverpa armigera*

Fig. 2.11.3. Larva of *Helicoverpa armigera* infected with *Nomuraea rileyi*

Common names: Cluster caterpillar or army worm
Scientific name: *Spodoptera litura* (F.)
Order: Lepidoptera
Family: Noctuidae
Distribution: South and Southeast Asia and South Pacific islands
Hosts: This is a polyphagous pest of crops belonging to the families Solanaceae, Malvaceae, Cruciferae, Fabaceae, Araceae, Alliaceae, Musaceae, and others.
Life cycle: Eggs are laid in clusters on leaves and are covered with hair scales for protection. Each egg cluster contains 100–300 eggs which hatch in 3 to 6 days. Larvae feed together as a cluster after hatching and disperse as they mature. There are six larval instars and, depending on the food plant and temperature, development to pupation takes 15–21 days. A full-grown larva is about 50 mm in length. Pupation occurs in an earthen cell in the ground. The pupal stage lasts about 12 days (Figs 2.11.4, 2.11.5, 2.11.6).
Damage: The most conspicuous damage is caused by early larval instars as hundreds of caterpillars feed in clusters and quickly skeletonize leaves.
Natural enemies: Numerous parasitoids and predators have been recorded on *S. litura*. Of these, the egg parasitoids, *Telenomus spodopterae* and *Telenomus remus*, the larval parasitoids, *Apanteles* spp., *Euplectrus* sp., *Palexorista* sp. and the endoparasitoid *Microplitis* sp., and predators such as ants, earwigs, pentatomid bugs, and predaceous beetles play an important role in reducing populations of *S. litura*. Other natural enemies include a nucleopolyhedrovirus (NPV) and a fungus, *Nomuraea rileyi*. The NPV is widely used for control of this pest in the field (Figs 2.2.7, 2.2.8).

Fig. 2.11.4. *Spodoptera litura* adult

Fig. 2.11.5. *Spodoptera litura* egg mass

Fig. 2.11.6. *Spodoptera litura* larva

Fig. 2.11.7. Grubs of the ectoparasitoid *Euplectrus* sp. on a larva of *Spodoptera litura*

Fig. 2.11.8. The tachinid *Palexorista* sp., a parasitoid of *Spodoptera litura*

Common name: Green tobacco capsid
Scientific name: *Nesidiocoris* (*Cyrtopeltis*) *tenuis* Reuter
Order: Hemiptera
Family: Miridae
Hosts: Tobacco, tomato
Distribution: South and Southeast Asia
Life cycle: Adults are 2.5–3.5 mm long. Eggs are laid under the surface of leaves or inserted into leaf veins. Nymphs feed at the site of hatching. There are four instars. The total development time from egg to adult is 26–33 days (Fig. 2.11.9).
Damage: This insect feeds on the stems of tomato blooms, causing a girdling effect that results in flower drop. Leaf damage has been reported in tobacco.
Natural enemies: Little is known about the natural enemies of this insect.
Other remarks: Like many mirids, this insect is both a plant feeder and a predator, feeding occasionally on small insects, eggs, young caterpillars, and nymphs of whiteflies.

Common and scientific names: (Leaf miners)
Stone leek leaf miner – *Liriomyza chinensis* (Kato)
Pea leaf miner – *Liriomyza huidobrensis* (Blanchard)
Serpentine leaf miner – *Liriomyza sativae* Blanchard
Chrysanthemum leaf miner – *Liriomyza trifolii* (Burgess)
Order: Diptera
Family: Agromyzidae
Distribution: *L. chinensis* – China, Japan, Indonesia
L. huidobrensis – Indonesia, China, Japan, Africa, Central and South America
L. sativae – India, Indonesia, China, Malaysia, Japan, Thailand, Vietnam, Africa
L. trifolii – India, the Philippines, Africa

Fig. 2.11.9. Adult *Nesidiocoris tenuis* feeding on a tomato stem

Hosts: *L. chinensis* – leek, onion, and shallot
L. huidobrensis – peas, beans, potato, tomato, and onion
L. sativae – polyphagous, tomato, cucurbits, crucifers, beans, and others
L. trifolii – polyphagous, tomato, cucurbits, crucifers, onions, beans, and others

Life cycle: The life cycle of the four species is similar. Females pierce the leaf surface with their ovipositor and feed on the exuding sap. Eggs are laid below the epidermis of the leaf. On emergence the young larvae mine the leaf tissue, the tunnels widen as the larvae grow. At the end of the third instar, the larvae cut a hole in the cuticle of the leaf and fall to the ground for pupation. The entire life cycle from egg to adult takes 3–9 weeks, depending on temperature. In Hawaii and Guam, accidental introduction of *L. trifolii* progressively displaced *L. sativae* as a major pest. *Liriomyza huidobrensis* prefers higher altitudes in the tropics (Fig. 2.11.10).

Fig. 2.11.10. *Liriomyza huidobrensis* adult

Fig. 2.11.11. Leaf miner damage to onion leaves

Fig. 2.11.12. Leaf miner damage and a pupa on a potato leaf

Damage: Tunneling of the larvae damages leaves resulting in reduced photosynthesis and yield. Heavy infestations can kill plants (Figs 2.11.11, 2.11.12).
Natural enemies: The parasitoids associated with *L. chinensis* in Japan include *Halticoptera ciculus, Pnigalio katonis, Diglyphus isaea, Chrysocharis pentheus, Neochrysocharis formosa,* and *Closterocerus trifasciatus. Neochrysocharis formosa* and *C. pentheus* were also found on *L. sativae* in Japan. *Neochrysocharis formosa* was the predominant parasitoid of *L. sativae* and *L. huidobrensis* in Liaoning Province of China. *Hemiptarsenus varicornis, Neochrysocharis* sp., and *Opius* sp. were found on *L. huidobrensis* in Indonesia. On tomatoes in greenhouses, *L. trifolii* is controlled by combined release of *Dacnusa sibirica* and *Diglyphus isaea* in Japan. Several species of parasitoids are known to attack these leaf miners in the U.S.

Common name: Western flower thrips
Scientific name: *Frankliniella occidentalis* (Pergande)
Order: Thysanoptera
Family: Thripidae
Distribution: North and South America, Europe, Hawaii, Malaysia, Réunion and Canary Islands, New Zealand, Uzbekistan
Hosts: Temperate to tropical crops including peaches, plums, nectarines, grapes, cotton, avocado, beans, lettuce, strawberry, alfalfa, mustard, onion, orange, spinach, capsicum, eggplant, cucumber, tomato, watermelon, and others
Life cycle: This insect is haplodiploid; unfertilized eggs develop into haploid males, while fertilized eggs develop into diploid females. Eggs are opaque, kidney shaped, and laid in the tissues of leaves, flowers or fruits. It passes through two nymphal stages, a prepupal stage and a pupal stage, before reaching adulthood. Newly emerged females are white to light yellow and later develop into pale, intermediate, or dark color forms. Males are uniformly pale. The reproductive rate is decreased at temperatures below 15°C and above 30°C.
Damage: This is a polyphagous pest and feeding by this insect causes disfiguring of young fruit in nectarines, flower drop in strawberries, and withering of leaves in cucumbers. This thrips is a vector of tospoviruses: *Chrysanthemum stem necrosis virus, Impatiens necrotic spot virus, Tomato chlorotic spot virus,* and *Tomato spotted wilt virus.* Only early instars can acquire the virus by feeding on infected plants and adults transmit it to healthy plants. Tomato spotted wilt virus infects tomato, pepper, tobacco, lettuce, peanut, and many ornamental crops.
Natural enemies: Recorded predators of this thrips include *Orius insidiosus* and *Orius tristicolor* in California, *Orius persequens* and *Paratriphleps laevisculus* in Hawaii, and *Nabis* spp. and *Geocoris* spp. in California. Other predators include lacewings, predaceous thrips, ladybeetles, syrphid flies, and mites. The parasitoids *Ceranisus americenis* and *Ceranisus russelli* and an unidentified chalcid were recorded in the USA (Waterhouse and Norris, 1989).

Common name: Solenopsis mealybug
Scientific name: *Phenacoccus solenopsis* Tinsley
Order: Hemiptera
Family: Pseudococcidae

Distribution: North America, South America, India, Israel, Pakistan, Thailand, Cambodia, China, Hawaii, Taiwan, West Africa
Hosts: Cotton, tomato, okra, and other vegetables, ornamentals, and weeds
Life cycle: Adult female is either viviparous or ovoviviparous. Eggs hatch within 0.5–2 hours. Nymphs have three instars in females and four instars in males. The developmental period from egg to adult is 3 weeks for males and 2 weeks for females. Males live for 1–2 days and the females live for 6–7 weeks.
Damage: This mealybug is a serious pest of cotton, and, during 2006–2008, it caused millions of dollars worth of damage to cotton in India and Pakistan (Wang *et al.*, 2010). Vegetable and ornamental crops were affected to a limited extent.
Natural enemies: In 2008, a parasitoid, *Aenasius bambawale,* possibly accidentally introduced with the mealybug, parasitized 20–70% of this pest. Since its discovery, this parasitoid has greatly reduced populations of the solenopsis mealybug wherever it occurs, so currently there is little need for control measures.

Common name: Striped mealybug, white tailed mealybug
Scientific name: *Ferrisia virgata* (Cockerell)
Order: Hemiptera
Family: Pseudococcidae
Distribution: Pantropical
Hosts: Tomato, eggplant, jute, cassava, coffee, cocoa, citrus, mango, papaya, pineapple, guava, potato, and *Leucaena* sp.
Life cycle: The female is elongate and oval, covered with powdery white wax and long thin waxy threads forming a distinct tail. Each female lays about 200–400 eggs. The nymphal stage is completed in 20–45 days, development rate depending on host plant and temperature. Nymphs remain close to adults resulting in clusters of mealybugs. Adult females live for 1–2 months and the males 1–3 days. Ants often tend this mealybug, feed on the copious amounts of honeydew produced and keep natural enemies away (Lapis, 1970; Awadallah *et al.*, 1979).
Damage: Heavy build-up of the population occurs on tender shoots, rachis, leaves, and fruits. This mealybug is reported to be a vector of cocoa swollen shoot viral disease in West Africa and citrus tristeza virus in Ghana.
Natural enemies: The parasitoids *Aenasius advena* and *Blepyrus insularis* were reported from India.

Other major pests are:
Asian papaya fruit fly, *Bactrocera papayae* (See Pests of Papaya)
Spider mites, *Tetranychus* spp. (See Pests of Eggplant)

3 Pests of Minor Vegetable Crops

3.1 Pests of Amaranths (*Amaranthus* spp., Amaranthaceae)

Typical vegetable amaranths originated from Southeast Asia. *Amaranthus tricolor* L. is by far the most common species present in the region. Like many indigenous plants, *A. tricolor* is considered a poor man's food, but in recent times this vegetable has found its way into supermarkets around the region. Many of the wild amaranths in the region are considered to be herbal plants and their medicinal properties are held in high regard. *Amaranthus tricolor* is a tall annual plant up to 2.5 m in the wild, but cultivated varieties are shorter. Many selections of amaranths from local landraces have been made. The key pests are often polyphagous and include *Bemisia* sp., *Cletus trigonus*, *Spoladea recurvalis*, and grasshoppers.

Common name: Beet webworm
Scientific name: *Spoladea recurvalis* (F.) (*Hymenia recurvalis*)
Order: Lepidoptera
Family: Pyralidae
Distribution: Worldwide, mostly in tropical and subtropical regions.
Hosts: Amaranth, beetroot, Chinese cabbage, spinach
Life cycle: The egg stage of *S. recurvalis* takes from 3 to 5 days. The caterpillar is smooth and light green and feeds on leaves. It resembles a leaf vein. The larval stages are completed in 10–15 days. The caterpillars are usually found on the lower surface of leaves, making clear patches with the top skin intact. Pupation occurs in a cocoon covered with soil. The pupal stage lasts 10–15 days.
Natural enemies: The larvae are parasitized by *Apanteles* sp. in India.

Common name: Slender rice bug
Scientific name: *Cletus trigonus* Thunberg
Order: Hemiptera
Family: Coreidae
Distribution: Asia
Hosts: Besides amaranths, this polyphagous insect has been reported on many crops (including kapok) and weeds.
Life cycle: Eggs are ruby-shaped, and the egg stage lasts 8 days. There are five nymphal instars that are completed in 12–22 days (Eamrucksa, 1987).
Damage: Feeding by nymphs and adults produces necrotic spots on the leaves.
Natural enemies: Natural mortality factors are not well understood for this common indigenous pest species.

Common names: Oriental migratory locust and short-horned grasshopper
Scientific names: *Locusta migratoria* L.
Atractomorpha sp.
Order: Orthoptera
Family: Acrididae
Distribution: Found all over tropical and subtropical Asia
Hosts: These are polyphagous insects and, in the instance of the locust, can cause significant defoliation when they occur in large numbers.
Life cycle: Eggs are laid in egg pods after the female grasshoppers have mated. In most short-horn grasshoppers, these are laid in the soil with the aid of the long ovipositor of the female. Upon emergence, the young grasshoppers, resembling adults without wings, feed on nearby vegetation and undergo about six stages before becoming adults. Population increase is primarily related to climatic factors. For *L. migratoria manilensis* (Meyen), eggs are laid in pods of 12–42 and placed inside soil. Eggs hatch in 17–22 days, and the nymphs develop into adults in 30–45 days (Kalshoven, 1981).
Damage: Heavy infestation by grasshoppers causes severe defoliation of the crop.
Natural enemies: In his review, Greathead (1992) suggested that natural biological control of grasshoppers provides a useful constraint to population increase, and it is likely that insect pathogens may be useful to suppress locust outbreaks. Some success with sprays of *Beauveria bassiana* in the Cape Verde Islands has been reported. Prior (1992) discovered a virulent strain of *Metarhizium* spp. that can control outbreaks of locusts.

Other major pests are:
Sweet potato whitefly, Cotton whitefly, Tobacco whitefly, *Bemisia tabaci* (Gennadius) (*B. argentifolii* or Biotype B – Silverleaf whitefly) (See Pests of Okra)
Corn earworm, cotton bollworm, tomato fruitworm, *Helicoverpa armigera* (See Pests of Tomato)
Cluster caterpillar or army worm, *Spodoptera litura* (See Pests of Tomato)

3.2 Pests of Beetroot (*Beta vulgaris* L., Chenopodiaceae)

Beetroot is of Mediterranean origin, and it is cultivated mostly as a vegetable crop at higher altitudes in tropical Asia.

Major pests include:
Green peach aphid, *Myzus persicae* (See Pests of Pepper)
Cluster caterpillar or army worm, *Spodoptera litura* (See Pests of Tomato)
Beet webworm, *Spoladea recurvalis* (*Hymenia recurvalis*) (See Pests of Amaranths)

3.3 Pests of Carrot (*Daucus carota* L., Apiaceae)

Carrot is a domesticated form of the wild carrot, *D. carota,* native to Europe and Southwest Asia. It has been selectively bred for its greatly enlarged and more palatable and edible taproot. In early use, carrots were grown for their aromatic leaves and seeds, not their roots. Some relatives of the carrot are still grown for these uses, such as parsley, fennel, dill, and cumin. In the tropics, carrot is grown in cooler upland areas at altitudes higher than 500 m. The most serious insect pests are those that attack the root, such as the carrot fly, wireworms, and grubs. Although many insects feed on the foliage, they usually do not cause serious damage due to the action of natural enemies.

Common name: Carrot fly
Scientific name: *Chamaepsila rosae* (F.)
Order: Diptera
Family: Psilidae
Distribution: Europe and Asia
Hosts: Carrot, parsnip, parsley, celery
Life cycle: Eggs are laid in the soil around the developing carrots. Upon hatching, larvae burrow into the surface of the root.
Damage: The maggots feed on the outer layers of the carrot root.
Natural enemies: Two fungal pathogens, *Entomophthora muscae* and *Entomophthora schizophorae,* have been reported infecting adult flies. There is evidence that another fungal pathogen, *Beauveria bassiana,* has potential for controlling the larval stage. A staphylinaid ectoparasite, *Aleochara* sp., was found parasitizing the puparia.

Common name: Green looper
Scientific name: *Thysanoplusia orichalcea* (F.)
Order: Lepidoptera
Family: Noctuidae
Distribution: Throughout Southeast Asia
Hosts: This looper can be found on a wide range of cultivated crops and wild plants.
Life cycle: Eggs are laid singly on leaves. The egg stage lasts 3–4 days. There are five larval instars lasting about 13 days. Pupation occurs in a silken cocoon attached to and covered by the carrot leaves.

Damage: Larvae feed on the leaves and occasionally they cause defoliation but usually not serious enough to reduce yields.

Natural enemies: Many larval parasitoids have been reared from field-collected loopers. There are also numerous predators such as carabids and pentatomids that feed on eggs and larvae. Insect pathogens play a major role in regulating looper populations, especially in carrots. Fungal pathogens such as *Nomuraea rileyi* and *Pandora gammae* are commonly found infecting larvae, and, if loopers are found in a carrot field, one can usually find the nucleopolyhedrovirus infecting that population. This is a phenomenon that the authors have observed in numerous upland locations throughout Southeast Asia.

Other remarks: There are two other species of loopers that may occur together with *T. orichalcea* in carrots, namely, *Chrysodeixis chalcites* (see Pests of Beans) and *Argyrogramma signata* (F.) (Lep. Noctinidae).

Other pests of regional importance are:
Pea leaf miner, *Phytomyza atricornis* (Diptera: Agromyzidae); sweet potato flea beetle, *Chaetocnema basalis* (Coleoptera: Chrysomelidae); wireworms (Coleoptera: Elateridae), and grubs (Coleoptera: Scarabeidae)

3.4 Pests of Kangkong, or Water Spinach (*Ipomoea aquatica* Forsk., Convolvulaceae)

Kangkong, also known in English as water spinach and water morning glory, is a semiaquatic tropical plant grown as a leafy vegetable. It is found throughout the world in tropical and subtropical areas but is most commonly grown in East and Southeast Asia. Its center of origin is Central to South China. Kangkong is used extensively in Malay, Indonesian, and Chinese cuisine. Kangkong is susceptible to damage by generalist foliage feeding insects such as leaf miners, caterpillars, leaf beetles, grasshoppers, whiteflies, and aphids. In general, the pest complex found on kangkong is similar to that found on sweet potato. However, very little information that is specific to kangkong is available concerning insect pests.

Major pests include:
Tortoise beetles, *Aspidomorpha miliaris* and others (See Pests of Sweet Potato)
Sweet potato hawkmoth, *Agrius convolvuli* (See Pests of Sweet Potato)
Sweet potato stem borer, *Omphisa anastomosalis* (See Pests of Sweet Potato)
Cluster caterpillar or army worm, *Spodoptera litura* (See Pests of Tomato)

4 Pests of Major Fruit Crops

4.1 Pests of Bananas (*Musa* spp., Musaceae)

Bananas and plantains, which evolved in Southeast Asia and India, respectively, are the most important fruit crops in the tropics. Bananas are grown in low input backyard gardens and in subsistence agriculture, as well as in high input export plantations. India is the world's largest producer of bananas, but bananas are also widely grown in South and Southeast Asia and the Pacific Islands. The banana weevil is one of the most serious pests of bananas throughout Asia, but the banana leaf roller and the scab moth are also common pests in Southeast Asia. Even though the direct damage caused by banana aphid is minimal, it is a serious pest because of its ability to vector banana bunchy top virus disease.

Common name: Banana weevil borer
Scientific name: *Cosmopolites sordidus* (Germar)
Order: Coleoptera
Family: Curculionidae
Distribution: This weevil is present in all banana growing areas of the world (Waterhouse and Sands, 2001)
Hosts: *Musa* spp. and *Ensete* spp. (Banana, plantain, and Manila hemp are severely affected)
Life cycle: Nocturnal adults live up to 2 years and lay up to 100 eggs per year. The eggs are white, ovoid, 2 mm long and are laid singly on the banana corm or pseudostem in small cavities gnawed out by the female near the ground surface. The egg stage lasts about 8 days, and larvae reach maturity in about 20 days. Larvae tunnel deep into the corm, return near the corm surface, and pupate in an oval chamber. The life cycle varies from 1 to 6 months depending on the temperature (Figs 4.1.1, 4.1.2, 4.1.3). Dispersal is mostly through infested plant material (Gold *et al.*, 2002).
Damage: Most damage is caused by larvae tunneling in the corm. Severe damage results in premature death of leaves, development of poor root systems, and toppling

Fig. 4.1.1. Adult of *Cosmopolites sordidus*

Fig. 4.1.2. Grub of *Cosmopolites sordidus*

Fig. 4.1.3. Pupal case of *Cosmopolites sordidus*

and breaking of the plants. Larval tunneling also may provide entry points for fungi and bacteria that cause rotting (Smith, 1997).

Natural enemies: The histerid beetle, *Plaesius javanus*, is commonly found preying on the banana weevil in Java and has been widely distributed to different parts of the tropical world since 1913. Another histerid beetle, *Plaesius laevigatus*, the staphylinids *Belonuchus ferrugatus* and *Priochirus unicolor*, a cucujid beetle, *Cathartus* sp., and the rhagionid fly *Chrysopilus ferruginosus*, are also known to attack *C. sordidus* larvae in Java. Also, a predatory hydrophilid beetle, *Dactylosternum hydrophiloides*, from Malaysia, a histerid, *Hololepta quadridentata*, from Trinidad, and the ant, *Tetramorium guineense*, in Cuba, are known to attack larvae of *C. sordidus* (Waterhouse, 1998; Waterhouse and Sands, 2001). Braithwaite (1958) reported a blue planarian worm, *Caenoplana coerulea*, attacking *C. sordidus* in Australia.

Common name: Pseudostem borer
Scientific name: *Odoiporus longicollis* (Olivier)
Order: Coleoptera
Family: Curculionidae
Distribution: India, Nepal, Myanmar, Sri Lanka, Thailand, Indonesia, and the Philippines
Hosts: Banana and manila hemp
Life cycle: Adults feed on live or dead banana leaf tissue but consume very little and are not generally considered pests. Adults are slightly larger than those of the banana weevil and readily fly. They live 6–24 months. Eggs are laid singly at the rate of one to six per day in the leaf sheaths in cavities made by the female rostrum. Each female is capable of laying 100–200 eggs. The egg stage lasts from 3–12 days. There are four or five instars and the larval stage lasts 2–10 weeks. The prepupal stage occurs in a pupal chamber made of leaf sheath fibers. Prepupal and pupal stages last 3–9 and 3–20 days, respectively.
Damage: The larvae bore through the inner leaf sheaths. Larval tunneling results in pseudostem rotting and in some cases the internal shoot is killed. Leaves of infested plants wither and fruits are undersized.
Natural enemies: No information is available.

Common name: Banana aphid
Scientific name: *Pentalonia nigronervosa* Coquerel
Order: Hemiptera
Family: Aphididae
Distribution: This aphid occurs in almost all countries where banana is grown.
Hosts: *Musa* spp. Foottit *et al.* (2010) have restricted *P. nigronervosa* to banana feeding populations and restored the name *Pentalonia caladii* van der Goot to the populations that feed on Zingiberaceae and Araceae.
Life cycle: Apterous and alate forms coexist. This aphid reproduces parthenogenetically throughout the year. Males have not been observed and reproduction is viviparous. There are four instars. Adults are about 1.5 mm in length and vary in color from reddish to dark brown. This pest completes 21–26 generations per year (Fig. 4.1.4).

Fig. 4.1.4 Banana aphid, *Pentalonia nigronervosa*

Damage: The preferred sites for feeding are within the whorl of the growing shoot, base of the leaves of young plants, and immature fruits. This aphid produces copious amounts of honeydew which attracts ants. This aphid transmits bunchy top virus which makes it a serious pest. Direct damage by the aphid is minimal. Bunchy top virus disease is established in Asia, some parts of Africa, the Mariana Islands, and Hawaii.
Natural enemies: Several parasitoids and generalist predators have been recorded for this aphid. Some of the parasitoids recorded are *Aphidius colemani*, *Aphidius* sp., *Lysiphlebus testaceipes*, and *Pseudendaphis maculans*. The predators recorded were *Cryptogonus orbiculus nigripennis*, *Micraspis discolor*, *Menochilus sexmaculatus*, *Coccinella septempunctata brucki*, and *Allograpta nasuta*. The fungi *Paecilomyces farinosus* and *Cephalosporium crassum* were also reported to infect this aphid (Waterhouse and Norris, 1987).

Common name: Banana scab moth
Scientific name: *Nacoleia* (*Lamprosema*) *octasema* (Meyrick)
Order: Lepidoptera
Family: Pyralidae
Distribution: Northeast Australia, Solomon Islands, Fiji, New Caledonia, Tonga, Samoa, Vanuatu, American Samoa, Wallis and Futuna Islands, New Guinea, Indonesia, and Malaysia (Paine, 1964; Gold *et al.*, 2002).
Hosts: *Musa* spp., *Heliconia* spp., *Nypa fruticans*, and *Pandanus* spp.
Life cycle: Eggs are oval, scalelike, and either laid singly or in small clusters of up to 20 overlapping eggs. Eggs are pale greenish white and are laid on the unopened inflorescences; they hatch in 4–6 days. Upon hatching, the young larvae crawl under the upper bracts of the bunch and feed on developing fruit. There are five larval instars (Hill, 1975). Pupation occurs within a silken cocoon covered with frass on the bunch, bracts, or dry leaf petioles. The life cycle is completed in about 28 days.
Damage: Larvae feed on the surface of the young developing banana fruits causing scarring and lesions that develop into black scabs. Severe damage may result in distorted to hollow fruits. Banana is the preferred host, but *Heliconia*, *Nipa*, and *Pandanus* serve as alternate hosts (Fig. 4.1.5).
Natural enemies: The parasitoids *Agathis* sp., *Apanteles* sp., *Macrocentrus* nr.

Fig. 4.1.5. Scab moth damage to banana bunch

trimaculatus, *Meteorus octasemae*, *Meteorus trichogrammae*, and *Chelonus striatigenas* were reported from Flores, and the latter was introduced from Flores to Java in 1959–60. An egg parasitoid, *Trichogramma* sp., was recorded from Malaysia. About 10 different parasitoids were recorded from the South Pacific Islands, but none were effective. The ant *Tetramorium bicarinatum* was observed attacking larvae in North Queensland (Waterhouse and Norris, 1987).

Common name: Banana skipper, banana leaf roller
Scientific name: *Erionota thrax* (Linnaeus)
Order: Lepidoptera
Family: Hesperidae
Distribution: Southeast Asia, Guam, Saipan, Hawaii, Papua New Guinea, Northeastern India, Andaman Islands, and Mauritius
Hosts: Bananas, *Musa textiles*, *Calamus trachycoleus*, *Caryota* sp., and nipa
Life cycle: The female butterfly lays eggs singly or in groups of about 25 mostly on the lower surface of the leaves. Eggs hatch in 5–8 days; first instars are pale green. Larvae move to the outer margin of the leaf where they start feeding and begin to roll the leaf. As the larvae grow, they cut and roll the leaf along the midrib. There are five instars, and all but the first instar are covered with white waxy powder. The larval stage lasts 23–30 days. Pupation occurs in the leaf roll and lasts from 8 to 12 days (Figs 4.1.6, 4.1.7, 4.1.8, 4.1.9).

Fig. 4.1.6. Butterfly of banana leaf roller, *Erionota thrax*

Fig. 4.1.7. Eggs of banana leaf roller, *Erionota thrax*

Fig. 4.1.8. Larva of banana leaf roller, *Erionota thrax*

Fig. 4.1.9. Larva starting the banana leaf roll

Fig. 4.1.10. Banana plant with leaf rolls caused by *Erionota thrax*

Fig. 4.1.11. Banana plants damaged by *Erionota thrax*

Damage: Heavy infestations may result in severe defoliation leaving only the leaf midrib with numerous rolls (Figs 4.1.10, 4.1.11).

Natural enemies: Egg parasitoids, *Agiommatus* sp. nr. *sumatraensis*, *Anastatus* sp., *Leurocerus hongkongensis*, *Ooencyrtus erionotae*, and *Ooencyrtus papilionis*; larval parasitoids, *Cotesia* (*Apanteles*) *erionotae*, *Bessa remota*, *Echthromorpha agrestoria*, *Pediobius* sp., *Scenocharops* sp., and *Sympiesis* sp.; and pupal parasitoids, *Brachymeria albotibialis*, *Brachymeria euploeae*, *Brachymeria lasus*, *Brachymeria marginata*, *Brachymeria* sp. nr. *marginicollis*, *Xanthopimpla regina*, and *Pediobius* sp., are effective in reducing populations of *E. thrax* (Waterhouse and Norris, 1989; Gold *et al.*, 2002). A larval parasitoid, *Elasmus philippinensis*, was reported on *Erionota torus* in Taiwan.

Common names: Rose beetles, chafer beetles
Scientific name: *Adoretus compressus* (Weber), *Adoretus sinicus* Burmeister
Order: Coleoptera
Family: Scarabaeidae
Distribution: Southeast Asia
Hosts: Polyphagous: banana, rose, grapes, okra, cotton, sweet potato, rambutan, coffee, cocoa, tea, oil palm, and others
Life cycle: Adults (10–12 mm long) lay eggs in the soil and developing grubs feed on decaying organic material. There are five instars, and pupation takes place in the soil.
Damage: The adults feed at night and can cause extensive damage by shredding margins of leaves. Because the beetles are nocturnal and hide during the day, farmers are often puzzled as to what is causing the damage.
Natural enemies: The parasitoids *Micromeriella* (*Campsomeris*) *marginella modesta* and *Tiphia segregata* were introduced to Hawaii and Guam to control this pest, but they were not effective.

Common names: Glasshouse thrips, black tea thrips
Scientific name: *Heliothrips haemorrhoidalis* (Bouché)
Order: Thysanoptera
Family: Thripidae
Distribution: Cosmopolitan
Hosts: Mango, banana, taro, rose, cocoa
Life cycle: Adults are 1.5–2.0 mm long and dark brown with greenish white legs and wings. Nymphs are greenish white. Reproduction is asexual, and males are not known to occur. Each female lays about 25 eggs. Immature stages often retain a globule of feces at the distal end of the abdomen. The life cycle is completed in 20–30 days (Fig. 4.1.12).

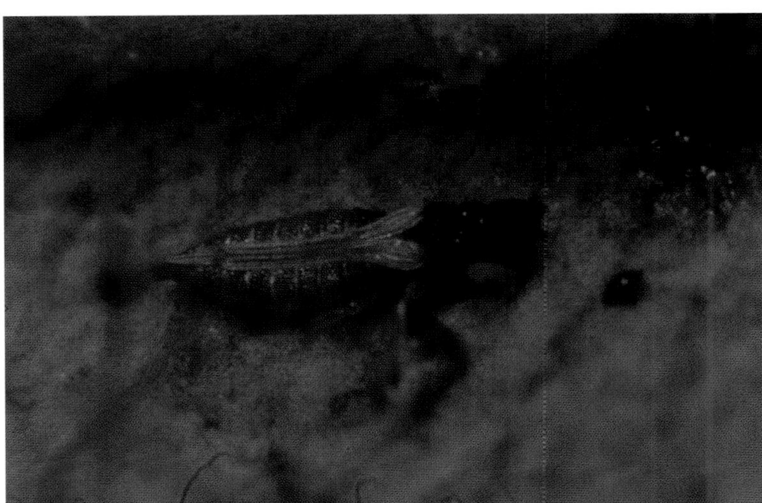

Fig. 4.1.12. Glasshouse thrips, *Heliothrips haemorrhoidalis*

Damage: The species is polyphagous and feeds mostly on the lower surface of leaves causing bleached patches covered with black spots (feces). Heavily infested plants become weak with discolored leaves.
Natural enemies: Heavy parasitism by *Thripobius javae* (=*Thripobius semiluteus*) and predation by the mite *Bdella distincta* on the larvae of *H. haemorrhoidalis* have been reported from Guam (Muniappan *et al*., 2000). The parasitoids *Goetheana shakespearei* (in Indonesia) and *Megaphragma mymaripenne* (in Hawaii) were reported from these thrips. *Thripobius javae* was imported from Italy into New Zealand and became established (Jamieson *et al*., 2008). Reported predators of these thrips include *Franklinothrips vespiformis* in California and Trinidad, *Franklinothrips tenuicornis* in Trinidad, and *Franklinothrips megalops* and *Chrysoperla carnea* in Israel (Watson *et al*., 1995).

Common name: Banana lace-bug
Scientific name: *Stephanitis typica* Distant
Order: Hemiptera
Family: Tingidae
Distribution: India, Palau, Thailand, Malaysia, Indonesia, Vietnam, the Philippines, Bangladesh, Myanmar, China, Hong Kong, Japan, Korea, Nepal, Sri Lanka, Maldives, Taiwan, and Papua New Guinea
Hosts: Banana, coconut, cardamom, turmeric, manila hemp, Alpinia
Life cycle: Eggs are embedded in the undersurface of leaves. Nymphs are gregarious and feed underneath leaves. There are five nymphal instars. Development from egg to adult takes about 33 days (Kalshoven, 1981). Adults (4 mm in length) are black and remain mostly on the lower surface of the leaves.
Damage: Feeding by this polyphagous species causes yellowish or whitish spots on the leaves.
Natural enemies: A predatory bug, *Stethoconus praefectus*, has been reported to feed on this bug. Other generalist predators almost certainly prey on the nymphs and adults.
Other major pests are:
Coconut scale, *Aspidiotus destructor* (See Pests of Papaya)
Spiraling whitefly, *Aleurodicus dispersus* (See Pests of Guava)
Nettle caterpillar, *Parasa philepida* (*lepida*) (See Pests of Beans)

4.2 Pests of Citrus (*Citrus* spp., Rutaceae)

Citrus originated in Southeast Asia in the region between northeast India and Yunan province in China. China, India, and Pakistan are the largest producers of citrus fruits in Asia. China is the world's largest producer of mandarins while India produces more lemons and limes. Citrus fruit is rich in vitamin C. Oranges, tangerines, grapefruit, and clementine are eaten fresh, and lemons and limes are used mainly as flavorings and juices. All citrus fruits can be made into juices and other value-added products and are ingredients in a variety of preparations. Major pests include citrus psyllid and citrus leaf miner (recent introductions to the New

World), whiteflies, scales, mealybugs, aphids, and fruit piercing moths. More pests on citrus have been reported in tropical than in subtropical regions (Smith and Peña, 2002).

Common names: Citrus psyllid, Asian citrus psyllid
Scientific name: *Diaphorina citri* Kuwayama
Order: Hemiptera
Family: Psyllidae
Distribution: All South and Southeast Asian countries, Florida, Hawaii
Hosts: *Citrus* spp., *Bergra koenigii, Murraya paniculata, Clausena anisumolens*
Life cycle: Eggs are laid singly inside half folded leaves, in buds, and in leaf axils. A female lays more than 800 eggs, and the egg stage lasts for 2–10 days depending on the temperature. Nymphs are flat and mostly found on young leaves (Fig. 4.2.1). Adults, which live up to 6 months, are 2.5 mm long and jump and fly when disturbed.
Damage: Adults and nymphs feed on young leaves and buds by sucking the sap, causing defoliation and wilting of shoots. The most damaging effect is the transmission of the gram-negative bacterium which causes citrus greening, known as huanglungbin in China.
Natural enemies: Parasitoids recorded are *Diaphorencyrtus aligarhensis* in the Philippines and Malaysia and *Tamarixia radiata* in Indonesia, Malaysia, Thailand, and Vietnam. *Tamarixia radiata* was introduced to the Philippines from Reunion in 1988. *Psyllaephagus* sp. was recorded in Indonesia (Kalshoven, 1981).

Fig. 4.2.1. Nymphs of the citrus psyllid *Diaphorina citri*

Common name: Citrus leaf miner
Scientific name: *Phyllocnistis citrella* Stainton
Order: Lepidoptera
Family: Gracillariidae
Distribution: Southern Europe, Asia, Africa, Australia, United States, Central America, Caribbean, Brazil, Chile, Colombia, Peru
Hosts: *Citrus* spp., mangosteen, *Eremocitrus glauca, Aegle marmelos, Jasminum humile*
Life cycle: Females lay single eggs on young leaves. Larvae bore into the leaf immediately after hatching. The fourth instar (prepupa) is about 3 mm in length (Fig. 4.2.2)
Damage: This leaf miner attacks young leaves, and, as a result, heaviest damage is seen in seedlings. Often, mined leaves are twisted, and the mines are silvery white in color (Fig. 4.2.2). This pest causes little damage above 1000 m in Indonesia. Incidence of citrus canker increases in leaf miner infested orchards.
Natural enemies: Parasitoids and predaceous arachnids, ants, and lacewings contribute to significant reduction of populations of this leaf miner. Some parasitoids recorded from the Southeast Asia region include: *Ageniaspis citricola* from Thailand, the Philippines, and Indonesia; and *Asecodes delucchi, Citrostichus phyllocnistoides, Closterocerus trifasciatus, Zaommomentedon brevipetiolatus, Sympiesis striatipes, Kratosyma citri*, and *Quadrastichus* sp. from Thailand.

Common name: Citrus mealybug
Scientific name: *Planococcus citri* (Risso)
Order: Hemiptera
Family: Pseudococcidae
Distribution: Cosmopolitan
Hosts: Wide host range including citrus, grapes, banana, coffee, fig, mango, cocoa, date, cassava, and passion fruit.
Life cycle: Adult females are yellow to yellowish brown and eggs are yellow. Females are oviparous and lay 300–500 eggs which hatch in 3–6 days.
Damage: This serious pest is known to transmit viruses of grapes, cocoa, cucumber, and taro.
Natural enemies: The coccinellid predators include: *Brumus suturalis, Scymnus apiciflavus, Scymnus roepkei*, and *Coccinella transversalis* (= *Coccinella repanda*); the

Fig. 4.2.2. Citrus leaf mines caused by *Phyllocnistis citrella*

cecidomyiid *Coccidodiplosis smithi*; and other generalists. Parasitoids recorded include *Leptomastix abyssinica* and *Anagyrus greeni*. The parasitoids *Prochiloneurus io, Prochiloneurus rex,* and *Holanusomyia pulchripennis* (from the Philippines) and *Tetracnemus indicus* (from India) also have been recorded. The fungus *Entomophthora fresenii* has been recorded from Indonesia.

Common name: Common mormon
Scientific names: *Papilio polytes* L. (Lime butterfly *Papilio demoleus* L., Blue mormon *Papilio polymnestor* Cramer)
Order: Lepidoptera
Family: Papilionidae
Distribution: Asia
Hosts: *Citrus* spp., *Murraya* spp., *Triphasia* sp.
Life cycle: Yellow eggs are laid singly on young shoots and they hatch in about 4 days. Young larvae are brown with white markings resembling bird droppings. Third instars and older caterpillars are green with black and white markings. The larval stages last 11–27 days and pupation takes place in a chrysalis attached to the host plant. The pupal stage is completed in 8–12 days. The adult is a black swallow-tailed butterfly with white markings (Figs 4.2.3, 4.2.4, 4.2.5, 4.2.6).

Fig. 4.2.3. The common mormon butterfly, *Papilio polytes*

Fig. 4.2.4. Early instar larva of *Papilio polytes*

Fig. 4.2.5. Later instar larva of *Papilio polytes*

Fig. 4.2.6. Chrysalis of *Papilio polytes*

Damage: Larval feeding can cause defoliation when populations are high. However, this rarely happens because of the action of natural enemies.
Natural enemies: *Trichogramma* sp. and *Telenomus* sp. on eggs; *Apanteles* sp., *Erycia nymphalidaephaga* Bar., *Charops* sp., and *Brachymeria* sp. on larvae; and *Pteromalus* sp. on pupae have been reported in Asia. Generalist predators also feed on eggs and young larvae.

Common name: Citrus fruit borer
Scientific name: *Citripestis sagittiferella* (Moore)
Order: Lepidoptera
Family: Pyralidae
Distribution: Thailand, Malaysia, Singapore, Brunei, Indonesia
Hosts: Citrus spp.
Life cycle: Eggs are laid in a mass on the undersurface of citrus fruits. The egg stage lasts 5–6 days. Larvae are reddish brown in early stages (Fig. 4.2.7) and turn green before pupation. Larval stages last about 9–13 days. Fully grown larvae leave the fruit and lower themselves to the ground using a silken thread. They pupate in the soil and adult moths emerge after 10 days.
Damage: Upon hatching, larvae bore through the rind, and then tunnel into the pulp and seeds. This causes premature fruit drop and fruit that is unmarketable.
Natural enemies: Larval parasitoids, *Atherigona orientalis, Cremastus* sp., and *Rhoptromeris* sp. from Malaysia and *Trichogrammatoidea* sp. nr. *nana* from Indonesia have been recorded.

Fig. 4.2.7. Larva of citrus fruit borer, *Citripestis sagittiferella*

Common name: Citrus rind borer
Scientific name: *Prays endocarpa* Meyrick
Order: Lepidoptera
Family: Yponomeutidae
Distribution: Malaysia, Singapore, Indonesia, the Philippines, French Guiana, Guadeloupe, Martinique, and Réunion
Hosts: Citrus spp.
Life cycle: The adult is a small moth with a wingspan of about 8 mm. Eggs are flat, transparent, and green. They are laid singly on the fruit. The larva eats its way through the bottom of the egg and bores straight into the rind. Larvae are initially green and later develop red transverse bands. The larval stages, which last for about 3 weeks, are spent in the rind. The feeding of the larva and lignification of the surrounding tissue result in formation of a swelling or gall. The mature larva leaves the gall and pupates on the fruit, stems, or edges of leaves. Pupation takes place in a loosely woven pale brown cocoon. The pupal period lasts 3–7 days. An adult moth lives up to 3 weeks and lays about 100 eggs during her lifetime.
Damage: The unsightly swellings on the fruit reduce market value. Little or no damage is done to the pulp, but premature fruit drop occurs in severe infestations.
Natural enemies: A larval parasitoid, *Euderus malayensis*, has been recorded from Malaysia.

Common name: Citrus bud borer
Scientific name: *Prays citri* Mill
Order: Lepidoptera
Family: Yponomeutidae
Distribution: Southern Europe to Southeast Asia
Hosts: Citrus spp.
Life cycle: This pest attacks flower buds of citrus but also can be found boring into young fruit and feeding on older fruit. No swellings are formed. Eggs are oval and caterpillars are yellowish or light green. Pupation takes place outside the feeding tunnel in a loosely woven cocoon.
Damage: This pest causes flowers and young fruit to drop.
Natural enemies: None reported.

Common name: Spined citrus fruit bug
Scientific name: *Rhynchocoris poseidon* Kirkaldy
Order: Hemiptera
Family: Pentatomidae
Distribution: Southeast Asia
Hosts: Citrus spp.
Life cycle: Eggs are placed in groups on leaves. Nymphs and adults suck juice from fruit. The fungus *Nematospora* sp. has been found to be closely associated with this bug in Indonesia, and the fungus is transmitted via the stylet while feeding. The life cycle of this bug is completed in 37–45 days (Figs 4.2.8, 4.2.9).
Damage: This is not a major pest; however, its feeding on developing fruit coupled with secondary infection by microorganisms can result in fruit drop.
Natural enemies: None reported.

Fig. 4.2.8. Spined citrus fruit bug, *Rhynchocoris poseidon*

Fig. 4.2.9. Spined citrus fruit bug *Rhynchocoris poseidon* feeding on a lemon

Common name: White moth cicada
Scientific name: *Lawana conspersa* (Walker)
Order: Hemiptera
Family: Flatidae
Distribution: Southeast Asia
Hosts: Citrus, coffee, durian, cocoa, and tea

Fig. 4.2.10. White moth cicada, *Lawana conspersa*

Fig. 4.2.11. White moth cicada, *Lawana conspersa* infected by a fungus

Life cycle: Eggs are inserted into young twigs or midribs of leaves. Nymphs, covered in white waxy material, feed on the shoot tips, young leaves, and flowers. They hop away when disturbed. Little information is available on the biology of this insect (Fig. 4.2.10).
Damage: Nymphal feeding may cause minor damage.
Natural enemies: The fungus *Metarhizium anisopliae* is known to attack this insect and keep it under control in Malaysia (Fig. 4.2.11).

Common name: Citrus blackfly
Scientific name: *Aleurocanthus woglumi* Ashby
Order: Hemiptera
Family: Aleyrodidae
Distribution: South and Southeast Asia, Central America, USA, Caribbean, East Africa, South Africa
Hosts: Citrus, coffee, guava, mango, cashew, and rose
Life cycle: Adults have a bluish pubescent body. Eggs are laid in a spiral on the lower side of leaves and hatch in about 9 days. Nymphs are sessile and black, but the third instar nymph has a whitish waxy band around the body. There are three instars and a pupal stage. Nymphal stages last from 45 to 115 days (Figs 4.2.12, 4.2.13, 4.2.14).
Damage: Nymphs suck sap and produce honeydew which encourages sooty mold to grow on the upper surface of the leaves. Fruit production is decreased, and tree vigor can be reduced by heavy infestations.
Natural enemies: The parasitoids *Eretmocerus serius*, *Encarsia smithi*, *Encarsia clypealis*, *Encarsia opulenta*, and *Amitus hesperidum* have been utilized for control of this pest in various parts of the world.

Fig. 4.2.12. Adult of *Aleurocanthus woglumi*

Fig. 4.2.13. Puparia of *Aleurocanthus woglumi*

Fig. 4.2.14. Eggs of *Aleurocanthus woglumi*

Common name: Orange spiny whitefly
Scientific name: *Aleurocanthus spiniferus* Quaintance
Order: Hemiptera
Family: Aleyrodidae
Distribution: Asia, Micronesian Islands, East Africa, and South Africa
Hosts: Citrus, rose
Life cycle: Similar to citrus blackfly
Damage: Similar to citrus blackfly
Natural enemies: *Encarsia smithi* has provided effective control of this whitefly in Guam, Kosrae, Pohnpei, Chuuk, and Yap islands in Micronesia. Introduction of *E. smithi* and *Amitus hesperidum* resulted in control of this pest in Hawaii.

Common names: Tea tortrix, Coffee leaf folder
Scientific name: *Homona coffearia* (Nietner)
Order: Lepidoptera
Family: Tortricidae
Distribution: Asia
Hosts: Citrus, coffee, tea, okra, rambutan, chiku, soybeans
Life cycle: Eggs are yellow, flat, and laid on the upper surface of tender leaves. The number of eggs laid by a female varies from 177 to 829. The incubation period ranges from 4 to 7 days, and there are six larval instars. A fully grown larva is 26 mm long. Larvae web young parts of plants, resulting in leaves being rolled up. The larval period ranges from 25 to 34 days, and pupation takes place within the folded leaf. The pupal period ranges from 4 to 8 days, and the total development time is 34–47 days (Figs 4.2.15, 4.2.16).
Damage: This pest may cause serious defoliation of host plants in some seasons. Besides leaves, larvae may also feed on flower buds and fruit.
Natural enemies: Many parasitoids are known to attack this insect in Java, Indonesia. In the Philippines, larval parasitoids, *Hemiptarsenus* sp., an unidentified bethylid, and a pupal parasitoid, *Brachymeria* sp., were recorded. Introduction of *Macrocentrus homonae* in 1936 resulted in successful biological control of this pest in tea in Sri Lanka.

Fig. 4.2.15. Adult of *Homona coffearia*

Fig. 4.2.16. Larva of *Homona coffearia*

Common name: Citrus bark borer
Scientific name: *Agrilus occipitalis* Esch.
Order: Coleoptera
Family: Buprestidae
Distribution: Philippines, Indonesia, and Papua New Guinea
Hosts: Citrus
Life cycle: The adult beetle is about 5 mm long and greenish brown in colour. It feeds on mature leaf margins causing saw-toothed edges. Eggs are laid on the bark of the mature stems, usually in lesions or cracks. The larvae bore into the inner bark tissue causing gummosis. Pupation takes place in a tunnel in the bark.
Damage: Tree vigor is lost as larvae bore into the tree bark, but adult feeding on leaves is less important. In severe infestations, the trees may die.
Natural enemies: No information is available.

Common name: Fruit piercing moth
Scientific name: *Eudocima* (*fullonia*) *phalonia* (L.).
Several species of fruit piercing moths are known to occur in Asia. In addition to the species mentioned above, other important ones include: *Eudocima materna* (Linnaeus), *Eudocima ancilla* Cramer, *Eudocima salaminia* (Cramer), *Platyja umminia* (Cramer), *Pericyma cruegeri* (Butler), *Ercheia dubia* (Butler), *Serrodes mediopallen*, and *Oraesia emarginata*.
Order: Lepidoptera
Family: Noctuidae
Distribution: Tropical Africa, Asia, Australia, and the Pacific
Hosts (larval): Vines of Menispermaceae and *Erythrina* spp.
Life cycle: Eggs are yellowish green and hemispherical; they are laid either singly or in groups on host plants of the family Menispermaceae in Asia, Australia, and Africa and on *Erythrina* spp. in the Pacific. Eggs hatch in 3–4 days, and the larvae pass through 5–6 instars. Early instars are dark brown, and later instars show different colorations varying from green to brown to black. The pupal stage lasts from 14 to 21 days. Adults exhibit sexual dimorphism. They fly long distances from their breeding grounds to fruit orchards during fruit bearing season (Figs 4.2.17, 4.2.18, 4.2.19, 4.2.20).
Damage: Larval feeding on menispermaceous vines and *Erythrina* foliage is not of economic significance. However, the adults pierce ripening or ripe fruit such as oranges, guava, pomegranate, mango, litchi, soursop, carambola, banana, and many others. Their highly chitinized proboscis is used to pierce fruits, macerating the tissue and sucking the resultant juice. They are active at night from 8 pm to 2 am. Several other secondary piercing moths feed on the fruits using holes caused by the primary piercers. Secondary microbial organisms enter the pierced fruits and cause rotting. Bagging individual fruits, netting the trees and orchards (as in the case of litchi orchards in Australia), and baiting are practiced to reduce damage by these pests.
Natural enemies: Egg parasitoids *Ooencyrtus* sp., *Telenomus* sp., and *Trichogramma* sp. and larval parasitoids *Euplectrus maternus*, *Euplectrus platyhypenae*, *Echthromorpha striata*, *Lissopimpla pacifica*, and *Winthemia caledoniae* have been recorded. Predators such as lygaeid and reduviid bugs, lacewings, and ants feed on eggs; mantids, pentatomids, and vespids prey on larvae.

Fig. 4.2.17. Fruit piercing moth, *Eudocima phalonia*, feeding on a citrus fruit

Fig. 4.2.18. Fruit piercing moth, *Eudocima phalonia*, feeding on a pomegranate fruit

Fig. 4.2.19. Moths feeding on guava fruit using the holes made by the fruit piercing moth

Fig. 4.2.20. Larva of the fruit piercing moth, *Eudocima phalonia*

Common names: Brown citrus aphid and Black citrus aphid
Scientific name: *Toxoptera citricidus* (Kirkaldy) and *Toxoptera aurantii* (Boyer de Fonscolombe)
Order: Hemiptera
Family: Aphididae
Host: Citrus
Distribution: Cosmopolitan
Life cycle: These aphids are parthenogenetic, viviparous, and lay about five young ones a day. There are four nymphal stages and it takes 1–6 weeks to complete the entire set of nymphal stages. Alate forms are produced when aphid colonies become dense.
Damage: Aphids cause direct damage by sucking sap. Also, *T. citricidus* is a vector of citrus tristeza virus disease.
Natural enemies: Parasitoids recorded include *Aphelinus gossypii, Aphidius colemani, Aphidius matricariae, Lysiphlebus testaceipes, Lysiphlebia japonica,* and *Lysiphlebus fabarum*. Lady beetles, syrphid fly larvae, and lacewing larvae are major predators.

Another major pest:
Fruit flies, *Bactrocera* spp. (See Pests of Mango)

Other minor pests: Several species of mites, mealybugs, scale insects, and others regionally important have been recorded. These include the citrus red mites *Phyllocoptruta oleivera* (Ashmead) and *Eriophyes sheldoni* (Ewing) (Acari: Eriophyidae); broad mite *Polyphagotarsonemus latus* (Banks) (Acari: Tarsonemidae); *Brevipalpus* spp. (Acari: Tenuipalpidae); *Panonychus citri* (McGregor), *Eutetranychus orientalis* Klein, *Eutetranychus cenadani* Rimando, and *Tetranychus urticae* Koch (Acari: Tetranychidae); *Aonidiella aurantii* (Maskell) and *Unaspis yanonensis* Kuwana (Hemiptera: Diaspididae); cottony cushion scale *Icerya purchasi* Maskell (Hemiptera: Margarodidae); *Ceroplastes* spp. (Hemiptera: Coccidae); *Darna trima* (Moore) (Lepidoptera: Limacodidae); and *Scirtothrips dorsalis* Hood and *Heliothrips haemorrhoidalis* (Bouché) (Thysanoptera: Thripidae).

4.3 Pests of Guava (*Psidium guajava* L., Myrtaceae)

Guava is native to tropical America, covering the region from Mexico to Peru and eastern Brazil. It is well adapted to hot climates and is grown throughout the subtropics and tropics. Commercially, guava is used to produce jams, jellies, and juices, and it may be eaten raw or cooked. It is rich in vitamin C, iron, calcium, and phosphorus. Fruit flies and fruit piercing moths are the serious constraints in guava production in the tropics.

Common name: Spiraling whitefly
Scientific name: *Aleurodicus dispersus* Russell
Order: Hemiptera
Family: Aleyrodidae

Distribution: Central and South America, Caribbean, the Pacific Islands, Asia, and Africa
Origin: Central America
Hosts: Polyphagous: This whitefly is a pest of vegetables, fruit trees, ornamentals, and shade trees.
Life cycle: Adults (2 mm) are white and lay eggs in spirals covered with waxy secretions, usually on the undersurface of leaves. The egg stage takes about 7 days. There are four larval instars which last about 30 days. Only the first instar is mobile. Adults and larvae feed on plant sap (Figs 4.3.1, 4.3.2, 4.3.3).

Fig. 4.3.1. Spiraling whitefly infestation on a guava leaf

Fig. 4.3.2. Adults of spiraling whitefly, *Aleurodicus dispersus*

Fig. 4.3.3. Waxy filaments produced by the spiraling whitefly on a guava leaf

Damage: Feeding by large populations can cause dessication of plants. Whiteflies also produce honeydew which encourages the growth of sooty mold. This blocks light that is necessary for normal plant growth.
Natural enemies: Introduction of the coccinellid *Nephaspis oculatus* and the parasitoid *Encarsia haitiensis* in combination with local natural enemies has resulted in reducing the pest status of the spiraling whitefly. Fortuitously, introduced parasitoids *Encarsia dispresa* and *Encarsia guadeloupe* keep this pest under control in West Africa (Obinna *et al.*, 2010).

Common names: Cocoa mirid, tea mosquito
Scientific name: *Helopeltis theivora* Waterhouse
Order: Hemiptera
Family: Miridae
Distribution: Asia
Hosts: Cocoa, mango, cashew, guava, tea, neem
Life cycle: The adult is slender, 6–8 mm long, agile, and a good flier. A female lays about 500 elongated, sausage shaped, white eggs that have two long, unequal, protruding filaments. Eggs are laid in buds and axils of leaves or in tender stems; they hatch into wingless nymphs in 5–7 days. Nymphs are greenish yellow and pass through 5 instars in 18 days. The adult is a black and red elongated insect with long antennae and legs. There is a dorsal process on the scutellum. Feeding habits of nymphs and adults are similar (Fig. 4.3.4).
Damage: Feeding punctures on shoots often result in dieback. On guava, flower buds and fruits are attacked.
Natural enemies: Biological control employing the ant, *Dolichoderus thoracicus*, is being considered in Malaysia and Indonesia.

Fig. 4.3.4. Adult tea mosquito, *Helopeltis* sp.

Common name: Atlas moth
Scientific name: *Attacus atlas* Linnaeus
Order: Lepidoptera
Family: Saturniidae
Distribution: Southeast Asia
Hosts: Avocado, guava, rambutan, custard apple, soursop, santol, citrus, banana
Life cycle: This is a large, reddish brown moth with a 28 cm wingspan. There are transparent "windows" on the wings. Globular, dirty brown eggs are laid singly on the undersurface of leaves. As these eggs tend to be laid on the same leaf, they appear to occur as a clump. Eggs hatch in 10–13 days. Larvae are white to pale green with thick, stubby spines. They feed on young and maturing leaves. The larval stage lasts 28–38 days. The full-grown larvae can grow to a length of 150 mm. Pupation takes place within a silken cocoon enclosed in a leaf. The pupal stage lasts 23–28 days.
Damage: Severe defoliation can result when the insect occurs in large numbers. This only happens occasionally because parasitoids and predators keep this pest under natural control.
Natural enemies: The parasitoid, *Tricholygia bombiscum,* is effective in keeping populations of this moth under control in the Philippines. Many general insect predators and insectivorous birds feed on the larvae.

Other major pests are:
Chili thrips, *Scirtothrips dorsalis* (See Pests of Pepper)
Red-banded thrips *Selenothrips rubrocinctus* (See Pests of Mango)
Oriental fruit fly and others, *Bactrocera dorsalis* (*Bactrocera zonata, Bactrocera frauenfeldi, Bactrocera cucurbitae*) (See Pests of Mango and Cucurbits)
Fruit piercing moth, *Eudocima phalonia* (See Pests of Citrus)
Striped mealybug, *Ferrisia virgata* (See Pests of Tomato)
Coffee mealybug, *Planococcus lilacinus* (See Pests of Soursop)
Citrus mealybug, *Planococcus citri* (See Pests of Citrus)
Coconut scale, *Aspidiotus destructor* (See Pests of Papaya)

4.4 Pests of Mango (*Mangifera indica* L., Anacardiaceae)

Mango originated in the region between India and the Philippines. Over the years mango cultivation has spread throughout tropical and subtropical regions of the world. It is a rich source of Vitamins A and C, beta-carotene, and fiber. India is one of the largest producers but accounts for only 1% of the international trade. Thailand, Pakistan, Bangladesh, and the Philippines are major exporters of mango in Asia. Quarantine restrictions on fruit flies and stone weevil are the major impediments to international trade of this fruit. Several species of fruit flies are known to infest mango in different parts of the world.

Common name: Mango tip moth
Scientific name: *Penicillaria jocosatrix* (Guenée)
Order: Lepidoptera
Family: Noctuidae
Distribution: South Asia, Southeast Asia, Australia, Hawaii, Guam, Palau
Hosts: Mango, Cashew, *Schinus* sp., *Spondias purpurea*, and *Terminalia* sp.
Life cycle: Eggs are laid singly on tender leaves and inflorescences. They are pale green when laid but turn brown the second day. The caterpillars are light green and covered with reddish dots. They turn pinkish brown before pupation occurs in a tough silken cocoon on the stem in the cracks of bark or in the soil (Fig. 4.4.1).
Damage: Larval feeding causes defoliation of the newly sprouted shoots and inflorescences. In the absence of effective natural enemies, this pest can result in total crop loss.
Natural enemies: In Malaysia, larval parasitoids *Euplectrus* sp. and *Apanteles* sp. have been recorded. *Blepharella lateralis* and *Euplectrus* sp. were introduced to Guam from India in 1986. *Euplectrus taiwanus* has been recorded from Taiwan.

Common name: Mango hoppers
Scientific names: *Idioscopus clypealis* Lethierry, *Idioscopus niveosparsus* Lethierry, *Idioscopus nitidulus* Walker, *Idioscopus incertus* Baker
Order: Hemiptera
Family: Cicadellidae
Distribution: Asia. *I. nitidulus* in Malaysia, *I. clypealis* in the Philippines, *I. niveosparsus* (Leth.) and *I. clypealis* (Leth.) in Indonesia, and *I. incertus* in Guangzi Province, China, are the common leafhopper pests on mango. *I. niveosparsus* was recorded in a small area in the northern part of Queensland, Australia.
Host: Mango
Life cycle: Adult females lay 100–200 eggs in their lifetime. Eggs are laid in rows of 2–12 and are inserted into the tender stems, midribs, and panicles. Eggs hatch in 4–7

Fig. 4.4.1. Larva of mango tip moth, *Penicillaria jocosatrix*

days and there are 4–5 instars. Nymphal stages last about 8–10 days, but the adults live up to 300 days. Mango hopper populations reach their peak during the flowering season (Figs 4.4.2, 4.4.3).

Damage: Feeding by adults and nymphs on flowers causes them to drop. Numerous oviposition punctures on the inflorescence also may contribute to flower drop. When there is a heavy loss of flowers, fruit set can be adversely affected. Mango hoppers produce honeydew which results in the growth of sooty mold that retards plant growth.

Natural enemies: An ectoparasitoid, *Epipyrops fuliginosa*, on adults has been recorded in India, Thailand, and Pakistan. The egg parasitoids *Gonatocerus* sp., *Mirufens* sp. nr. *mangiferae*, *Centrodora* sp. nr. *scolypopae*, *Aprostocetus* sp., and *Quadrastichus* sp. in Pakistan and *Aprostocetus* sp., *Gonatocerus* sp., and *Polynaema* sp. in India have been reported (Waite, 2002). The egg parasitoid, *Centrodora* sp., was common and parasitized 40% of the eggs in Indonesia. In addition, several predator species, such as lacewings, mantids, and spiders, have been reported feeding on the nymphs and adults (Kalshoven, 1981). In Bangladesh, two fungi, *Aspergillus origer* and *Fusarium* sp., were found to infect adult hoppers.

Common name: Red banded mango caterpillar
Scientific name: *Deanolis sublimbalis* Snellen (*Noorda albizonalis* Hampson, *Autocharis albizonalis* (Hampson))
Order: Lepidoptera
Family: Pyralidae
Distribution: India, Myanmar, Thailand, China, the Philippines, Indonesia, Brunei, Papua New Guinea
Hosts: Mango
Life cycle: Eggs are white to crimson and are laid in groups of 1–4 on or near the peduncle. Eggs hatch in 3–4 days, and larvae have 5 instars in 14–20 days. Larvae have a brown or black head and white body with a red ring in each abdominal

Fig. 4.4.2. Adult mango hopper, *Idioscopus nitidulus* (courtesy of Ho Cheng Tuck)

Fig. 4.4.3. Mango inflorescence damaged by *Idioscopus nitidulus*

segment. They first feed on the pulp of the fruit and later bore into the seed. Pupation takes place in an earthen cocoon in the soil. The pupal period lasts 9–14 days, and adult longevity is 8–9 days.
Damage: Mango fruit is attacked at all developmental stages. The caterpillar feeds on fruit as small as 2 cm and destroys the fruit from the inside out. Fruits do not reach maturity when attacked by this caterpillar. It is a serious pest of mangoes in South and Southeast Asia and Papua New Guinea (AQIS, 2002).
Natural enemies: A trichogrammatid parasitoid from eggs and a vespid larval predator have been recorded in the Philippines.

Common name: Red-banded thrips
Scientific name: *Selenothrips rubrocinctus* (Giard)
Order: Thysanoptera
Family: Thripidae
Distribution: Asia, Australia, Africa, South America, Caribbean, Central America, Florida
Hosts: Polyphagous; some important hosts are: mango, avocado, cashew, guava, and cocoa.
Life cycle: Adults are black, and, although females are common, males are rare or absent. Reproduction is probably parthenogenetic. Eggs are inserted into leaf tissue next to a vein. Nymphs are yellow with a characteristic red band on each of the first three abdominal segments. A drop of liquid excrement is carried by nymphs on the tip of their upward curled abdomen. When this falls on the leaf surface, it produces a dark spot. The larval stage lasts 4–8 days, prepupa 1–4 days, and pupa 4–7 days. The total life cycle requires 28–43 days.
Damage: Both adults and nymphs feed on the foliage and fruit causing scarification. Infested leaves have a silvery sheen. Heavy infestation reduces fruit production.
Natural enemies: Weaver ants, *Oecophylla smaragdina,* in Australia and *Orius* sp.in South Africa were reported to feed on red-banded thrips. The parasitoid, *Goetheana* (*Dasyscapus*) *parvipennis,* was introduced from Ghana to the West Indies between 1933 and 1937 to control *S. rubrocinctus*. It has become established in Trinidad, Jamaica, and Puerto Rico (Callan, 1943) where it parasitized 100% of the *S. rubrocinctus* population in greenhouses but only 20% to 30% in the field (Lewis, 1997).

Common name: Mango shoot borer
Scientific name: *Chlumetia transversa* (Walker)
Order: Lepidoptera
Family: Noctuidae
Distribution: Southeast Asia
Host: Mango
Life cycle: The adult is a gray moth with coffee-brown spots on the wings. Eggs are laid singly on tender leaves. The newly hatched larva bores into the midrib, and after a day or two it moves to the terminal shoot. As the larva develops, it tunnels downwards causing the shoot to wither and die. The mature larva leaves the shoot and pupates on the bark of the tree or in the soil. The entire life cycle takes about 22–27 days (Figs 4.4.4, 4.4.5).

Fig. 4.4.4. Damage to mango shoot by the mango shoot borer *Chlumetia transversa*

Fig. 4.4.5. Mango shoot borer damage

Damage: Larval feeding causes shoots to die which leads to yield reduction.
Natural enemies: A larval parasitoid, *Megaselia chlumetiae,* has been reported from India.

Common name: Mango stone weevil
Scientific name: *Sternochetus mangiferae* (F.)
Order: Coleoptera
Family: Curculionidae
Distribution: Asia (absent in the Philippines), Africa, the Pacific islands
Host: Mango
Life cycle: Adults live from 1 to 2 years and diapause most of the time except during fruiting season. Adults lay eggs on developing fruit by cutting a small boat-shaped incision. The exudate from the wound covers the egg and provides a protective opaque coating. One female lays about 15 eggs per day and a total of 300 eggs. Eggs hatch in 5–7 days, and larvae have 5–7 instars. The larval period ranges from 22 days to 10 weeks, and the pupal stage lasts about 1 week. Time required from egg to adult is 35–54 days, and there is only one generation per year. Adults diapause in crevices in the bark or under stones (Fig. 4.4.6).
Damage: This insect is monophagous and is a pest of quarantine importance. Seeds of infested fruits are still capable of germination, especially polyembryonic ones. Over 70% of damaged single-seeded cultivars also can germinate. This indicates that concerns about the negative impact of the weevil are exaggerated (Follett and Gabbard, 2000). There is limited economic damage as the larva feeds mainly within the stone and rarely in the pulp.
Natural enemies: The generalist predator ant, *Oecophylla smaragdina,* was reported to feed on adults.

Fig. 4.4.6. Mango stone weevil, *Sternochetus mangiferae*

Common name: Mango pulp weevil
Scientific name: *Sternochetus frigidus* (F.)
Order: Coleoptera
Family: Curculionidae
Distribution: Malaysia, Palawan Island in the Philippines, parts of Indonesia
Host: Mango
Life cycle: Eggs are laid on young fruit, and the larvae burrow into the flesh forming brown chambers. They pupate within the fruit in small cavities. Adult weevils emerge from the fruit by making small exit holes.
Damage: The mango pulp weevil looks similar to the mango stone weevil, but the pulp weevil only attacks flesh causing it to rot (AQIS, 2002).There are no signs of entry on the exterior of the fruit.
Natural enemies: None reported.

Common names: Oriental and other fruit flies
Scientific names: *Bactrocera dorsalis* (Hendel) [*Bactrocera zonata* (Saunders), *Bactrocera frauenfeldi* (Hardy)]
Order: Diptera
Family: Tephritidae
Distribution: Asia
Hosts: Mango and many other fruits and vegetables.
Life cycle: Females insert their eggs directly into the host fruit. There are three larval instars, and a prepupal stage followed by pupation within a cocoon usually found in the soil below the host plant (Waterhouse, 1993a). The complete life cycle varies from 18 to 29 days (Yunus and Balasubramaniam, 1981), the egg stage lasting 1–2 days; 9–18 days are required for larval development, and the pupal period lasts from 8 to 10 days (Figs 4.4.7, 4.4.8).

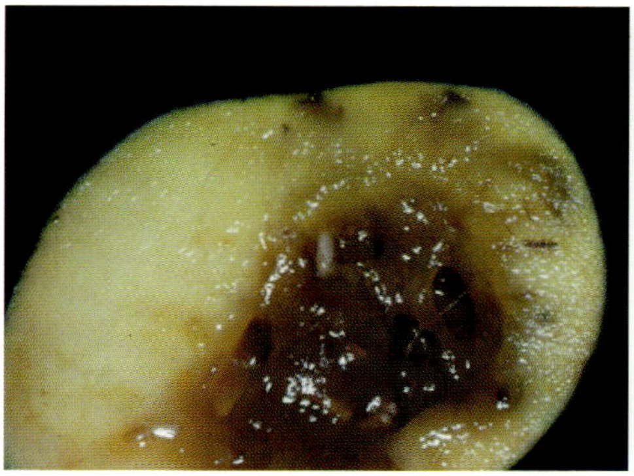

Fig. 4.4.7. Mango fruit damaged with fruit fly larvae

Fig. 4.4.8. Oriental fruit fly, *Bactrocera dorsalis*

Damage: Female flies often sting immature fruits without laying eggs. This causes a sap exudate that discolors the skin of the fruit. Eggs are laid in fruits close to maturity and larvae tunnel through the flesh causing the fruit to rot (Waite, 2002). Fruit flies are a serious constraint to the export of mangoes.
Natural enemies: Parasitoids *Aceratoneuromyia indica, Diachasmimorpha hageni Diachasmimorpha longicaudata, Fopius arisanus, Fopius vandenboschi, Psyttalia incisi, Tetrastichus dacicida,* and *Tetrastichus giffardianus* have been recorded (Waterhouse, 1993a).

Other remarks: The common name, oriental fruit fly (*B. dorsalis*), was used until the early 1990s when it was reported as one of a complex of 52 species that occur in Asia. *Bactrocera carambolae, Bactrocera occipitalis, Bactrocera philippinesis, Bactrocera infestans,* and *Bactrocera papayae* are other species that have been reared from mango fruits.

Common name: Gold dust weevil
Scientific name: *Hypomeces squamosus* (F.)
Order: Coleoptera
Family: Curculionidae
Distribution: Southeast Asia
Hosts: Mango, guava, sapota, jackfruit, durian, rambutan, citrus, coffee, cocoa, and long bean
Life cycle: Adults are about 14 mm long. Eggs of the weevil are laid singly in the soil. Larvae develop in the roots of weeds or crop plants. Pupation occurs in the soil. Details of the life cycle are still unknown (Fig. 4.4.9).
Damage: Adult weevils feed on foliage and may cause considerable defoliation.
Natural enemies: Little is known about natural enemies of this pest. A tachinid fly has been reported to attack the larval stage.

Fig. 4.4.9. Gold dust weevil, *Hypomeces squamosus*

Common name: Mango leaf cutting weevil
Scientific name: *Deporaus marginatus* Pascoe
Order: Coleoptera
Family: Curculionidae
Distribution: Malaysia, Thailand, Myanmar, Singapore
Life cycle: A slit is made along the leaf midrib into which eggs are inserted singly. The leaf is cut off near its base, and further development of the egg occurs within the excised leaf. Eggs hatch in about 2 days. Upon hatching, the larvae mine the fallen leaf, resulting in irregular blotch mines. There are three instars that last for about 1 week. Full-grown larvae emerge from the mines and pupate in the soil. The pupal stage lasts about 9 days (Figs 4.4.10, 4.4.11, 4.4.12, 4.4.13, 4.4.14).
Damage: Adults feed on young leaves, and the ovipositing females cause defoliation. Tender shoots that are attacked dry up, and the growing point produces new shoots. Severe attack will affect flower production.
Natural enemies: No information is available on natural enemies of this pest.

Common name: Mango trunk and branch borer
Scientific name: *Rhytidodera simulans* White
Order: Coleoptera
Family: Cerambycidae
Distribution: Southeast Asia
Hosts: Mango, cashew, rose apple, star fruit
Life cycle: Eggs are deposited on the bark of branches, and, upon hatching, larvae bore into twigs. As they develop, they move into the larger twigs, branches, and trunk. Before pupating, mature larvae prepare an exit hole which is plugged with wood fiber. Adults are nocturnal and live for 50–100 days. Each female lays about 160 eggs. The life cycle takes 7–8 months.

Fig. 4.4.10. Adult *Deporaus marginatus*

Fig. 4.4.11. Eggs of *Deporaus marginatus*

Fig. 4.4.12. Larva of *Deporaus marginatus*

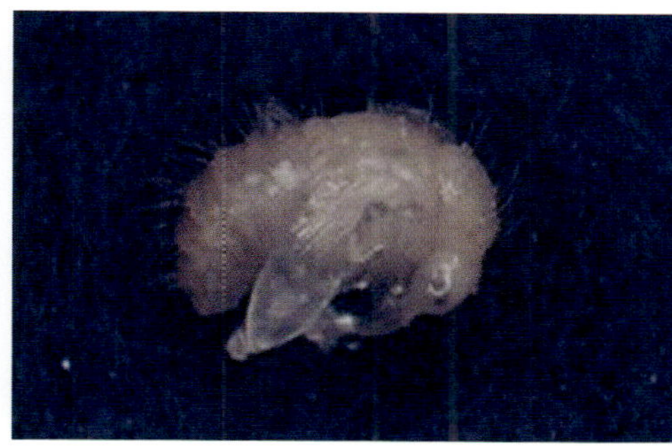

Fig. 4.4.13. Pupa of *Deporaus marginatus*

Fig. 4.4.14. Damage to mango leaf by *deporaus marginatus* adults

Damage: This pest causes drying of branches and, in severe infestations, kills the whole tree.
Natural enemies: No information is available on natural enemies of this pest.

Common name: Mango webworms
Scientific names: *Orthaga exvinacea* H. (in India), *Orthaga icarusalis* (Walker) (in Malaysia)
Order: Lepidoptera
Family: Pyralidae
Distribution: India, Malaysia, possibly most countries in South and Southeast Asia
Hosts: Mango, durian, cashew
Life cycle: The larvae live inside clusters of webbed leaves. Full-grown larvae are about 28 mm in length and reddish brown with a black lateral stripe on each side. Pupation takes place in the web in a cocoon with frass scattered on the outside. Little is known about the life cycle of this pest.
Damage: The larvae feed on the tender leaves in the shoots by webbing them together with silken threads. Serious defoliation and delayed flowering can result.
Natural enemies: No information is available.

Common name: The baron
Scientific name: *Euthalia aconthea* (Cramer)
Order: Lepidoptera
Family: Nymphalidae
Distribution: Malaysia, Thailand, Indonesia
Hosts: Mango, cashew
Life cycle: Eggs are laid singly on the young leaves. Caterpillars are light green and possess a single row of long, horizontally projecting, featherlike spines on each side of their bodies. A dorsal yellow stripe is present. The green chrysalis is rhomboid in shape with decorative spots. The chrysalis is attached to leaves or twigs (Fig. 4.4.15).
Damage: This is a minor pest of mango. Sometimes outbreaks occur on cashew.
Natural enemies: No information is available.

Fig. 4.4.15. Larva of *Euthalia aconthea* on a mango leaf

Common name: Mango mealybugs
Scientific names: *Rastrococcus spinosus* (Robinson); *Rastrococcus invadens* Williams
Order: Hemiptera
Family: Pseudococcidae
Distribution: *R. spinosus* - Malaysia; *R. invadens* - India, Indonesia, and Africa.
Hosts: Mango, citrus, rambutan, and cocoa
Life cycle: There are three nymphal stages, and adult males have one pair of wings, well-developed eyes, antennae, and legs. Females are wingless and covered with long waxy filaments (Figs 4.4.16a, b).
Damage: Feeding of the mealybugs on leaves, coupled with sooty mold development due to the production of honeydew, can affect the growth of the trees.
Natural enemies: Parasitoids and predators keep these pests under control in Indonesia and Malaysia. These scales are minor pests with only occasional outbreaks. The parasitoids *Gyranusoidea tebygi* and *Anagyrus mangicola* were introduced to West Africa from India for control of *R. invadens* in 1987 (Fig. 5.4.17). Ladybird beetles of the genus *Scymnus* and 7 species of parasitoids have been recorded on *R. spinosus* in Malaysia.

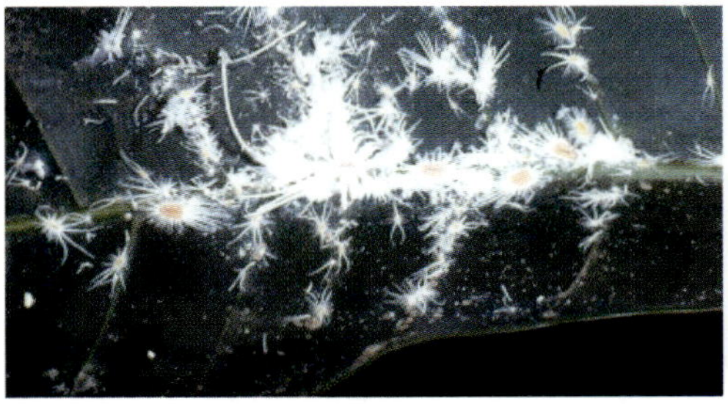

Fig. 4.4.16a Mango mealybugs *Rastrococcus spinosus* on a mango leaf

Fig. 4.4.16b Mango mealybugs on a mango leaf

Fig. 4.4.17. The parasitoid *Gyranusoidea* sp. ovipositing in a mango mealybug

Common name: Rose Coreid
Scientific name: *Mictis longicornis* Westwood
Order: Hemiptera
Family: Coreidae
Distribution: Southeast Asia
Hosts: Mango, cocoa, and jackfruit
Life cycle: Eggs are laid in a single line of about 6 eggs. Both adults and nymphs feed on leaves and inflorescences. Little is known about the biology of this insect.
Damage: This bug is a serious pest of young mango leaves and inflorescences in Malaysia. Affected young shoots and inflorescences wilt, leading to failure of fruit set.
Natural enemies: An egg parasitoid, *Chrysochalcissa oviceps*, has been recorded in Malaysia.

Other major pests are:
Fruit piercing moth, *Eudocima phalonia* (See Pests of Citrus)
Glasshouse thrips, *Heliothrips haemorrhoidalis* (See Pests of Bananas)
Coconut scale, *Aspidiotus destructor* (See Pests of Papaya)

Other pests of regional importance:
Mango gall midge, *Procontarinia mangicola* (Shi) (Dip.: Cecidomyiidae); mango fruit borer, *Citripestis eutraphera* (Lep.: Pyralidae); mango stem borer, *Batocera rufomaculata* DeGeer (Col.: Cerambycidae); mango gall psyllid, *Apsylla cistella* (Buckton) (Hem.: Psyllidae); and several scales and mealybugs.

4.5 Pests of Papaya (*Carica papaya* L., Caricaceae)

Papaya is of tropical American origin and is grown throughout the tropics. Papaya is cultivated for the fresh fruit market, green fruits that are used in cooking (especially in Asia), and papain collection. Fruit flies are constraints in export trade and require disinfestation treatment. Papaya mealybug originated in Mexico; it has spread progressively to the Caribbean, tropical South America, the Pacific, Asia, and, recently, Africa, causing devastation of papaya cultivation. However, implementation of a classical biological control program has effectively controlled this pest in countries where it invaded. Aphids, *Aphis gossypii* and *Myzus persicae*, are important as vectors of papaya ring spot virus.

Common name: Papaya mealybug
Scientific name: *Paracoccus marginatus* Williams and Granara de Willink
Order: Hemiptera
Family: Pseudococcidae
Distribution: Caribbean, Central and South America, Florida, Guam, Palau, Tinian, Hawaii, Bangladesh, Thailand, Malaysia, India, Indonesia, Cambodia, Philippines, Pakistan, Oman, Réunion Island, Ghana, Benin, Nigeria
Hosts: This pest has a wide host range of over 60 species of plants including economically important ones such as *Annona squamosa, Carica papaya, Hibiscus rosa-sinensis, Ipomoea* sp., *Manihot esculenta, Plumeria* sp., and *Solanum melongena*.
Life cycle: A detailed study of the life cycle of the papaya mealybug is lacking. Adult females are yellow and covered by a white waxy secretion. They lay 100–600 eggs in

an ovisac. The eggs hatch in 10 days, and the crawlers actively seek new sites for feeding. Females have four instars while the males have five. Males are winged, and females are sessile. A generation takes approximately 1 month. In general, populations of mealybugs build up in the dry season and decline in the rainy season (Figs 4.5.1, 4.5.2, 4.5.3).

Fig. 4.5.1. Papaya fruits infested with papaya mealybug and sooty mold

Fig. 4.5.2. Papaya tree affected by papaya mealybug with premature drying and curling of leaves

Fig. 4.5.3. Papaya trees killed by the papaya mealybug in Indonesia (Courtesy of Aunu Rauf)

Damage: Heavy infestations occur along the midrib and veins in older leaves and in all areas of tender leaves and fruit. While feeding, the mealybug injects a toxin into the plant. This, plus feeding activity, results in chlorosis, stunting, deformity, and early leaf and fruit drop. Honeydew produced by this mealybug causes sooty mold on leaves and fruits, thereby reducing photosynthetic surface. Heavy infestations kill papaya trees if no control measures are implemented.
Natural enemies: Five species of parasitoids, *Apoanagyrus californicus, Anagyrus loecki, Acerophagus papayae, Pseudaphycus* sp., and *Pseudleptomastix mexicana,* were collected from *P. marginatus* in Mexico. Of these, *A. loecki, A. papayae,* and *P. mexicana* were introduced to various countries for biological control of *P. marginatus.* The coccinellid beetle, *Cryptolaemus montrouzieri,* is a common predator on this mealybug.

Common name: Asian papaya fruit fly
Scientific name: *Bactrocera papayae* Drew and Hancock
Order: Diptera
Family: Tephritidae
Hosts: Papaya, citrus, banana, guava, *Syzygium,* apple, mango, carambola, cashew, all other edible fruits and fleshy vegetables
Distribution: Thailand, Malaysia, Singapore, Indonesia, Papua New Guinea
Life cycle: This is a polyphagous pest that can lay eggs in green papaya, young banana, and citrus. Females have a long ovipositor that allows them to penetrate beyond the sap layer of green fruit. The egg stage lasts for 2–3 days, and maggots develop through 3 instars inside the fruit, feeding on the pulp. Mature larvae exit the fruit and pupate in the soil within a puparium. The pupal stage lasts for about 10 days (Fig. 4.5.4).
Damage: Fruit that is attacked by this fruit fly will rot. It is a serious quarantine pest. This fly was found in Cairns, Northern Australia in 1992; it has since been eradicated by using male annihilation and protein bait spraying, at an expense of AUD 35 million.

Fig. 4.5.4. The fruit fly, *Bactrocera papayae*

Natural enemies: Some of the parasitoids recorded in Southeast Asia are *Biosteres persulcatus, Fopius arisanus, Diachasmimorpha longicaudata, Fopius vandenboschi, Psyttalia incisi,* and *Psyttalia* (*Opius*) *fletcheri.* The Asian weaver ant, *Oecophylla smaragdina,* is known to prey on fruit fly larvae.

Other remarks:
Other fruit flies that attack papaya in Asia are *Bactrocera dorsalis, Bactrocera cucurbitae, Bactrocera melanotus,* and *Bactrocera trilineola.*

Common name: Coconut scale
Scientific name: *Aspidiotus destructor* Signoret
Order: Hemiptera
Family: Diaspididae
Distribution: Throughout the tropics
Hosts: This scale is a pest of many tropical crops including coconut, mango, banana, avocado, breadfruit, cassava, cotton, guava, and papaya.
Life cycle: Adult females are circular, sessile, and covered by a translucent, flexible scale which is flat, whitish, and waxy. Males have a pair of wings, well-developed eyes, legs, and antennae, but no mouthparts. Eggs are laid under the scale at an average of 90 per female. Eggs hatch into nymphs, known as crawlers, in about 8 days. These crawl and settle on a suitable spot, insert their mouthparts, and begin feeding. They secrete wax which covers them. They develop and molt within this waxy shield. The period from egg to adult is about 32 days (Fig. 4.5.5)

Fig. 4.5.5. Papaya fruit covered with the scale insect, *Aspidiotus destructor*

Damage: In severe infestations, and in the absence of natural enemies, this scale can kill coconut palms. The removal of sap by the sucking mouthparts and toxicity of saliva result in leaves turning yellow and eventually drying up.
Natural enemies: Classical biological control programs have been successfully adopted in several countries to control the coconut scale. Several hymenopterous parasitoids and coccinellid beetles are known to attack this scale insect. Natural enemies utilized include *Aleurodothrips fasciapennis, Aphytis chrysomphali, Aphytis lingnanensis, Bardylis parvipennis, Chilocorus cacti, Chilocorus malasiae, Chilocorus nigritus, Chilocorus politus, Comperiella bifasciata, Comperiella unifasciata, Cryptogonus orbiculus nigripennis, Cryptognatha gemellata, Cryptognatha nodiceps, Cryptolaemus montrouzieri, Delphastus* spp., *Encarsia citrina, Pentilia insidiosa, Pseudoazya trinitalis, Pseudoscymnus anomalus, Rhyzobius satelles, Scymnus* sp., *Spaniopterus crucifer, Telsimia nitida, Zagloba aeneipennis,* and *Zaomma microphagum.*

Other major pests are:
Cotton aphid or Melon aphid, *Aphis gossypii* (See Pests of Eggplant)
Green peach aphid, *Myzus persicae* (See Pests of Pepper)
Spiraling whitefly, *Aleurodicus dispersus* (See Pests of Guava)
Spider mites, *Tetranychus* spp. (See Pests of Eggplant)
Broad mite or Yellow tea mite, *Polyphagotarsonemus latus* (See Pests of Pepper)

Other pests of regional importance are:
Palm scale, *Aspidiotus orientalis* Newstead, Oriental scale, *Aonidiella orientalis* (Newstead) (Hem.: Diaspididae), Red and black flat mite, *Brevipalpus phoenicis* (Geijskes) (Acari: Tenuipalpidae), and *Thrips parvispinus* Karny (Thysanoptera: Thripidae).

4.6 Pests of Pineapple (*Ananas comosus* Merr., Bromeliaceae)

The pineapple is indigenous to South America but is widely grown on a commercial scale in Asia. Pineapple ranks third after banana and mango in fruit crops harvested in the tropics and subtropics (Petty *et al.*, 2002). It is marketed as fresh fruit, canned, and made into juice, jam, and other products. Thailand, India, Indonesia, and the Philippines are the major producers of pineapple in Asia. The pineapple mealybug is by far the most important pest of this fruit crop. Petty *et al.* (2002) have reviewed pests of pineapple on a global scale.

Common name: Pineapple mealybug
Scientific name: *Dysmicoccus brevipes* (Cockerell)
Order: Hemiptera
Family: Pseudococcidae
Distribution: Widespread wherever pineapple is grown
Hosts: The preferred host is pineapple, but this pest also occurs on soybean, coffee, rice, and other crops.

Life cycle: The adult female is pinkish with a thick waxy covering and short, conical, waxy projections. It is parthenogenetic and ovoviviparous. About 250 crawlers are produced by a female. The female has three nymphal instars and the male has two. Both have prepupal and pupal stages. The adult male lives 1–3 days and the female 17–49 days (Fig.4.6.1).
Damage: This mealybug transmits a virus that causes stunting, reddening, and wilting known as pineapple mealybug wilt.
Natural enemies: The parasitoid, *Anagyrus ananatis*, effectively controls this mealybug in the absence of ants (Gonzalez-Hernandez, 1995). It was introduced to Hawaii from Brazil in 1934–1935 (Waterhouse, 1998). Other parasitoids introduced to Hawaii are *Euryrhopalus propinquus, Hambletonia pseudococcina, Leptomastix dactylopii, Pseudaphycus dysmicocci,* and *Pseudaphycus* sp. The coccinellids, *Scymnus apiciflavus* and *Sticholotis quatrosignata,* and seven species of unidentified hymenopteran parasitoids have been recorded in Malaysia. *Cryptolaemus montrouzieri* and *Rhyzobius ventralis* have been reported from the Pacific.

Common name: Pineapple scale
Scientific name: *Diaspis bromeliae* (Kerner)
Order: Hemiptera
Family: Diaspididae
Distribution: Occurs wherever pineapple is grown
Hosts: Agave, pineapple, bromeliads
Life cycle: Eggs are laid within the scale cover, and the egg stage lasts for 7 days. There are three larval stages and a pupal stage. It takes about 2 months to complete the life cycle.
Damage: Pineapple scale mostly infests lower leaves. Ratoon crops are more subjected to this scale attack as these plants are denser. A yellow spot develops at the place of feeding of the scale due to the injection of toxic saliva. In heavy infestations, fruit cracking may occur.

Fig. 4.6.1. Pineapple mealybug, *Dysmicoccus brevipes*

Natural enemies: The parasitoids, *Ablerus elegantulus*, *Encarsia* sp., and *Coccidencyrtus ochraceipes* (in South Africa), and *Aphytis* sp. and *Aspidiotiphagus* sp. (in Australia) have been recorded. The ladybird beetle, *Rhyzobius lophanthae*, is an important predator of this scale (Petty *et al*., 2002).

Common name: Pineapple fruit mite
Scientific name: *Steneotarsonemus ananas* (Tyron)
Order: Acari
Family: Tarsonemidae
Distribution: Wherever pineapple is grown
Host: Pineapple
Life cycle: Female mites lay oval and opaque white eggs singly. Larvae are white, and adults are creamy brown. The life cycle is completed in 1–2 weeks.
Damage: These mites feed on basal leaf tissue, flower bracts, and sepals, causing light brown spots (Paull and Duarte, 2011).
Natural enemies: A predatory thrips, *Podothrips lucasseni*, has been recorded.

Common names: Pineapple flat mite, pineapple red mite
Scientific name: *Dolichotetranychus floridanus* (Banks)
Order: Acari
Family: Tenuipalpidae
Distribution: Wherever pineapple is grown
Hosts: Pineapple, cardamom, grass, bamboo
Life cycle: Eggs are light orange, larvae are amber, nymphs are orange, and adults are red. The life cycle is completed in 10 days.
Damage: These mites cause severe damage when fruits mature under drought conditions. They feed on the basal tissue of leaves, especially of the crown (Paull and Duarte, 2011).
Natural enemies: *Amblyseius benjamini* is a common predatory mite recorded on *D. floridanus*.

Other major pests are:
Oriental fruit fly, *Bactrocera dorsalis* (See Pests of Mango)
Onion thrips, *Thrips tabaci* (See Pests of Onions)

5 Pests of Minor Fruit Crops

5.1 Pests of Avocado (*Persea americana* Mill., Lauraceae)

The avocado is of Central American origin but it has been transported to many tropical countries. Although it performs well in the subtropics, it is also grown widely in the tropics. Like many exotic crops, the avocado in its new environment is relatively free from insect pests. Details of the common pests in the area near its origin are provided by Wysoki *et al.* (2002). Most pests of avocado in California are under satisfactory biological control. The polyphagous insects that cause harm to the avocado plant include: *Attacus atlas, Valanga nigricornis, Ferrisia virgata, Aleurodicus dispersus*, and dacini fruit flies.

Common names: Javanese grasshopper or Malaysian locust
Scientific name: *Valanga nigrocornis* (Burmeister)
Order: Orthoptera
Family: Acrididae
Origin: Southeast Asia
Distribution: Southeast Asia
Hosts: It is polyphagous; in addition to avocado, it attacks coconut, rubber, and oil palm.
Life cycle: Eggs are laid in pods containing 70–100 in the soil. Each brown pod is 1–3 cm long (Ooi, 1988; Khoo *et al.*, 1991). The egg stage lasts about two months. Nymphs are greenish, and there are six instars in males and seven in females. Nymphal stages last for 80 days, and the life cycle is completed in about 6.5–8.5 months.
Damage: Nymphs and adults cause severe defoliation in certain seasons in Southeast Asia.
Natural enemies: Egg pods are attacked by ants and predatory beetles (Kalshoven, 1981). Eggs are parasitized by *Scelio* spp. During wet weather, the muscardine fungus, *Metarhizium* sp., infects nymphs.

Other major pests are:
Atlas moth, *Attacus atlas* (See Pests of Guava)
Spiraling whitefly, *Aleurodicus dispersus* (See Pests of Guava)
Striped mealybug, *Ferrisia virgata* (See Pests of Tomato)
Oriental fruit fly, *Bactrocera dorsalis* (See Pests of Mango)
Coconut scale, *Aspidiotus destructor* (See Pests of Papaya)
Red-banded thrips, *Selenothrips rubrocinctus* (See Pests of Mango)
Greenhouse thrips, Black tea thrips, *Heliothrips haemorrhoidalis* (See Pests of Bananas)
Tea red spider mite, *Oligonychus coffeae* (See Pests of Tea)
Indian wax scale (*Ceroplastes ceriferus*), Florida wax scale (*Ceroplastes floridensis*), Pink wax scale (*Ceroplastes rubens*) (See Pests of Caimito)

5.2 Pests of Breadfruit (*Artocarpus altilis* (Parkinson) Fosberg, Moraceae)

Breadfruit is a large fast-growing tree in the mulberry family. It originated in New Guinea and, in ancient times, it was spread by Polynesians to most Pacific Ocean islands and by Indonesians through Southeast Asia. More recently, it has been spread to most tropical regions of the world. The name is derived from the texture and flavor of the cooked fruit which are similar to fresh-baked bread. The fruit is rich in starch and, before eaten, is roasted, baked, fried, or boiled. It is a staple food in many tropical regions. It appears that relatively few pests attack this fruit crop.

Common name: Breadfruit stem borer
Scientific name: *Xyleborus* sp.
Order: Coleoptera
Family: Scolytidae
Distribution: Different species in this genus are found throughout Southeast Asia
Hosts: Citrus, jackfruit, fig, mango, pear, other fruits
Life cycle: Female beetles bore into the bark, forming a gallery. They deposit eggs in the gallery. Larvae form their own galleries as they feed and pupate at the end of these galleries. Emerging beetles bore out through the bark.
Damage: The beetles bore small holes about the size of pencil lead through the bark of twigs causing wilting and yellowing of foliage. Heavy infestations by this borer can kill young trees.
Natural enemies: None reported.

Common name: Breadfruit mealybug
Scientific name: *Icerya aegyptiaca* (Douglas)
Order: Hemiptera
Family: Monophlebidae
Distribution: Throughout Asia, also in Africa and Australia
Hosts: Jackfruit, citrus, fig, mango, pear, other fruits

Fig. 5.2.1. Breadfruit mealybug, *Icerya aegyptiaca*

Life cycle: Sessile females lay 300–400 eggs each. Nymphs become full-grown in 6 weeks, and the life cycle is completed in about 40 days. This scale multiplies rapidly and can cause heavy infestations. It is more abundant during drought conditions (Fig. 5.2.1).
Damage: Leaves turn yellow and drop prematurely, heavily infested shoots die, and yield is reduced up to 50%. Large amounts of honeydew promote sooty mold growth.
Natural enemies: The predatory beetle, *Rodolia cardinalis*, is an effective predator on this scale insect. Also, there are parasitoids of this species in India and Pakistan which may prove useful for biocontrol programs.

Common name: Breadfruit fly
Scientific name: *Bactrocera umbrosa* (Fabricius)
Order: Diptera
Family: Tephritidae
Distribution: Southeast Asia
Hosts: Breadfruit and jackfruit
Life cycle: A female lays 200–400 eggs which hatch in the fruit in 2 days. The larval stage lasts 6–9 days in the fruit. Larvae emerge from the fruit and pupate in the soil for 8–11 days. The adult life span is 8–14 weeks.
Damage: Feeding by the numerous larvae can destroy large portions of the fruit.
Natural enemies: A braconid wasp, *Phaenocarpa (Asobara) bactrocerae*, was recorded from the Philippines. Parasitoids recorded on other species of fruit flies may also attack this fruit fly.

5.3 Pests of Caimito, Cainito, or Star Apple (*Chrysophyllum cainito* L., Sapotaceae)

This fruit tree is a native of Central America and the Caribbean. It was recently introduced to the Philippines and Vietnam where its fruit is commonly served as a dessert. The caimito fruit borer is the most common pest.

Common name: Caimito fruit borer
Scientific name: *Alophia* sp.
Order: Lepidoptera
Family: Pyralidae
Host: Caimito
Distribution: Philippines
Life cycle: Eggs are laid singly on maturing fruits. The caterpillar bores into the fruit. No information is available on its biology.
Damage: Mature fruit is favored by this pest. It can damage 30% of the fruit in some seasons (Cendaña *et al.*, 1984).
Natural enemies: No information is available.

Common names: Indian wax scale (*C. ceriferus*), Florida wax scale (*C. floridensis*), Pink wax scale (*C. rubens*)
Scientific names: *Ceroplastes ceriferus* (F.)
Ceroplastes floridensis Comstock
Ceroplastes rubens Maskell
Order: Hemiptera
Family: Coccidae
Distribution: Old and New World; distribution varies depending upon species
Life cycle: Female scales covered with thick wax lay 1000–2000 eggs under them. Crawlers emerge from the eggs and move around. After molting, they insert their stylets into the plant and remain attached, covering themselves with wax (Fig. 5.3.1).

Fig. 5.3.1. Pink wax scale, *Ceroplastes rubens*

Damage: These scales mostly occur on stems and branches; they cause dieback and wilting of stems and reduced vigor. Severe infestations disfigure plants and scales produce copious amounts of honeydew on which sooty mold develops.
Natural enemies: The parasitoids *Scutellista caerulea, Coccidoctonus dubius, Metaphycus* sp., *Microterys australicus, Moranila californica,* and *Moranila compere* have been recorded on *C. ceriferus*. A ladybeetle, *Cryptolaemus montrouzeri,* and larvae of the moth *Mataeomera dubia* feed on crawlers and mature stages of *C. floridensis,* respectively. The parasitoids *S. caerulea, M. californica, Coccophagus ceroplastae, Diversinervis elegans, Microterys neitneri,* and *Aprostocetus ceroplastae* were recorded on *C. floridensis* in Australia. However, introduction of *Anicetus beneficus* from Japan into Australia in 1976 provided effective control of *C. rubens* (Waterhouse and Sands, 2001).

Other major pests are:
Oriental fruit fly, *Bactrocera dorsalis* (See Pests of Mango)
Coffee mealybug, *Planococcus lilacinus* (See Pests of Soursop)

Other minor pests are: green scale, *Coccus viridis* Green and hemispherical scale, *Saissetia coffeae* (Walker) (Hem.: Coccidae).

5.4 Pests of Carambola or Star Fruit (*Averrhoa carambola* L., Oxalidaceae)

Carambola originates from South and Southeast Asia, and it is widely cultivated in Asia and Central and South America. It is produced mainly for local and export fresh fruit markets. Malaysia is one of the major producers in Asia. Major pests are fruit flies and fruit piercing moths.

Major pests are:
Fruit piercing moth, *Eudocima phalonia* (See Pests of Citrus)
Oriental fruit fly and others, *Bactrocera dorsalis* (*Bactrocera zonata, Bactrocera frauenfeldi, Bactrocera cucurbitae*) (See Pests of Mango and Cucurbits)

5.5 Pests of Durian (*Durio zibethinus* Murr., Bombacaceae)

The durian fruit tree is a native to Brunei, Indonesia, and Malaysia, but it has been introduced into Australia and several countries in Asia. Thailand is the major exporter of durian, followed by Malaysia and Indonesia. In the Philippines, it is produced in Mindanao. China is the major importer of this fruit. It has a distinct flavor and an odor prompting people to express diverse views. Durian is used in a variety of culinary dishes and folk medicines in Southeast Asia. Major pests that attack this fruit tree are yellow peach moth and durian fruit borer.

Common name: Yellow peach moth
Scientific name: *Conogetes punctiferalis* (Guenée) (*Dichocrocis punctiferalis*)
Order: Lepidoptera
Family: Pyralidae
Distribution: Asia
Hosts: Durian, castor, rambutan, ginger, cocoa
Life cycle: Eggs are laid singly on the fruit. The larva has a black head and a pale greenish body with a pinkish suffusion dorsally. This coloration is known to vary with type of food. When full-grown, the larva measures about 20 mm in length. Pupation takes place in the tunnel within a silken cocoon. Pupal duration is about 8 days (Figs 5.5.1, 5.5.2).
Damage: The larva bores into the husk of the fruit, forming galleries. Damage is isolated and husks are frequently not bored through. However, it predisposes the fruit to disease infection.
Natural enemies: A pupal parasitoid, *Brachymeria* sp., has been recorded in Malaysia.

Common name: Durian fruit borer
Scientific name: *Mudaria magniplaga* (Walker) (*Plagidicta magniplaga*)
Order: Lepidoptera
Family: Noctuidae
Distribution: Thailand, Malaysia
Host: Durian
Life cycle: Eggs are laid on young durian fruit. Immediately after hatching, the larvae bore into the fruit. The larvae are pink with a reddish brown head and a dark brown prothoracic shield. Full-grown larvae are about 35 mm in length. Pupation takes place in the soil. The pupal period varies from 6 to 10 months.
Damage: This moth is a major pest of durian and causes substantial loss of marketable fruit. Often the whole fruit is inedible as a result of infestation by this pest.
Natural enemies: None recorded.

Fig. 5.5.1. Larva of yellow peach moth, *Conogetes punctiferalis*, on durian fruit

Fig. 5.5.2. Pupa of yellow peach moth in durian fruit

Common name: Gelatine grub
Scientific name: *Chalcocelis albiguttatus* (Snellen)
Order: Lepidoptera
Family: Limacodidae
Distribution: Malaysia
Hosts: Durian, rambutan, Malay apple, coffee, tea, coconut
Life cycle: Eggs are flat, scalelike, and laid on the undersurface of the leaves. Incubation period lasts for 2–3 days. Caterpillar stage lasts 59–70 days. As the common name suggests, the caterpillar appears to be enclosed in a gelatinous cover. Pupation takes place in the leaves inside a cocoon formed by the full-grown caterpillar. Pupal period lasts 21–31 days (Fig. 5.5.3).
Damage: Occasional outbreaks occur, and the caterpillars cause moderate to severe defoliation.
Natural enemies: Several natural enemies have been recorded in Malaysia. Natural enemies and microbial diseases keep the population of this pest under control.

Other major pests are:
Cocoa bark borer, *Squamura disciplaga* (See Pests of Rambutan)
Cotton leaf roller, *Sylepta derogata* (See Pests of Okra)
Mango webworms, *Orthaga exvinacea* (India), *Orthaga icarusalis* (Malaysia) (See Pests of Mango)
White moth cicada, *Lawana conspersa* (See Pests of Citrus)
Gold dust weevil, *Hypomeces squamosus* (See Pests of Mango)

Fig. 5.5.3. Gelatine grub, *Chalcocelis albiguttatus*

5.6 Pests of Jackfruit (*Artocarpus heterophyllus* Lamk., Moraceae)

Jackfruit is a large tree belonging to the mulberry family. It is believed to be indigenous to the southwestern rain forests of India, but is widely grown in South and Southeast Asia, East Africa, Brazil, and the Caribbean. Its fruit, which is borne on the trunk and main branches, is the largest tree-borne fruit, reaching 36 kg in weight and up to 90 cm in length. Its flesh is starchy, fibrous, and similar in flavor to a tart banana.

Common names: Jackfruit leaf-webber, banyon tussock moth
Scientific name: *Perina nuda* F.
Order: Lepidoptera
Family: Lymantriidae
Distribution: South and Southeast Asia
Hosts: Jackfruit, fig, mango, banyon
Life cycle: This pest is more abundant during the rainy season. The female moth lays 50–400 eggs in clusters or rows on the leaves. Eggs are light pink when freshly laid and turn red as they develop. Larvae in South and Southeast Asia go through six molts in 16–20 days. Pupation is in pupal cases on the upper surface of leaves. Moths emerge from pupae in 5–9 days and the entire life cycle takes 27–39 days.
Damage: Caterpillars fold or web leaves together and feed on them. They can cause extensive defoliation.
Natural enemies: Chalcid parasitoids include *Brachymeria croceogastralis*, *B. euploeae*, and *B. lasus*. A picorna type virus (PnPV) and a nucleopolyhedro virus (PnMNPV) are effective pathogens of this pest and can be used as microbial control agents (Cho *et al*., 1996).

Other major pests are:
Pink waxy scale, *Ceroplastes rubens* (See Pests of Caimito)
White tailed mealybug, striped mealybug, *Ferrisia virgata* (See Pests of Tomato)

5.7 Pests of Lanzones or Langsat (*Lansium domesticum* Correa, Meliaceae)

The lanzones are native to peninsular Malaysia and is grown in Malaysia, Indonesia, the Philippines, Cambodia, Thailand, and India. It is one of the favorite fruits in the Philippines. A bark borer and fruit flies are the most serious pests of this fruit tree.

Common name: Lanzones bark borer
Scientific name: *Prasinoxena* sp.
Order: Lepidoptera
Family: Galeriidae

Distribution: Philippines
Life cycle: This small green moth lays eggs in the cracks in the bark of lanzones trees. The larvae bore into the living tissue of the bark. Feeding of the larvae causes the bark to loosen and peel off. Larvae are brownish pink and are about 12 mm in length when fully grown. Pupation takes place under the dead bark (Cendaña *et al.*, 1984).
Damage: This is one of the most damaging insects on lanzones. Heavy infestations result in weakening of the trees and reduced fruit production.
Natural enemies: No information is available.

Other major pests are:
Coffee mealybug, *Planococcus lilacinus* (See Pests of Soursop)
Oriental fruit fly, *Bactrocera dorsalis* (See Pests of Mango)

Other pests of minor importance are Mango twigcutter *Niphonoclea albata* Newman (Col.: Cerambycidae) Bagworm *Eumata fuscescens* (Snellen) (Lep.: Psychidae).

5.8 Pests of Litchi (*Litchi chinensis* Sonn., Sapindaceae)

The litchi is a subtropical tree native to Southern China. It is an evergreen tree reaching 10–20 m. It is grown in Thailand, Taiwan, Japan, Vietnam, India, Australia, South Africa, and Brazil. The fruit is marketed fresh, dry, canned, or as juice (Paull and Duarte, 2011). The key arthropod pests are litchi stinkbug, fruit piercing moths, and litchi eriophyid mite.

Common name: Litchi stinkbug
Scientific name: *Tessaratoma papillosa* (Drury)
Order: Hemiptera
Family: Tessaratomidae
Distribution: China, Malaysia, Indonesia, India, Myanmar, the Philippines, and Vietnam
Hosts: Litchi, longan, citrus, pomelo, castor, pomegranate, eucalyptus, canna, loquat, and rose (Waite and Hwang, 2002)
Life cycle: The egg stage takes about 13 days. The five nymphal stages are completed in about 80 days. This stinkbug feeds on shoot terminals, flowers, and fruits.
Damage: Nymphs and adults pierce and feed on the flowering and fruiting shoots, causing flowers and fruits to drop. The feeding also causes necrosis of young twigs and blackening of fruit exocarp.
Natural enemies: In its center of origin, the litchi stinkbug is often under biological control; Waite and Hwang (2002) reported that the egg parasitoids *Encyrtus* (*Ooencyrtus*) sp., *Anastatus* sp., and *Blastophaga* sp. are key mortality factors in Guangdong. Liu *et al.* (2000) reported that eggs were attacked by *Anastatus japonicus* and *Ooencyrtus phongi*. In Thailand, Nanta *et al.* (1995) reported that four releases of 50,000 egg parasitoids per acre at intervals of 2 weeks during the critical period of severe outbreak from early February to March can achieve 70–80% control of this

pest. This would save longan growers about U.S.$600,000 in cost of pesticide applications over a year. To complement the parasitoids of *T. papillosa*, Lin (2005) suggested that effective strains of *Beauveria bassiana* could be applied.

Common name: Litchi eriophyid mite
Scientific name: *Aceria litchii* (Keiffer)
Order: Acarina
Family: Eriophyidae
Distribution: China, India, Taiwan, Pakistan
Life cycle: The female mites lay single eggs in the erinia produced on the leaf surface by their feeding. Eggs are minute, spherical, and translucent white. The life cycle is completed within the erinia. This mite is small, long, pinkish white, and has only four legs.
Damage: Feeding by this mite on young leaves results in formation of feltlike erinia on the underside of the leaf surface. Leaves are ruined in severe infestations and in some cases flowers and fruits are also affected.
Natural enemies: Many phytoseid predatory mites have been recorded feeding on this mite (Wu *et al.*, 1991; Waite and Gerson, 1994). *Agistemus exsertus* has been used for control of this eriophyid mite in China.

Other major pests are:
Striped mealybug, *Ferrisia virgata* (See Pests of Tomato)
Fruit piercing moth, *Eudocima phalonia* (*Eudocima materna, Eudocima ancilla, Eudocima salaminia, Platyja umminia, Pericyma cruegeri, Ercheia dubia, Serrodes mediopallens,* and *Oraesia emarginata*) (See Pests of Citrus)
Rose beetle, chafer beetle, *Adoretus compressus, Adoretus sinicus* (See Pests of Banana)

Other pests of regional importance are fruit borer, *Conopomorpha sinensis* Bradley, in China and Thailand, leaf-miner, *Conopomorpha litchiella* Bradley, in Thailand, shot hole borer, *Xyleborus fornicatus* (Eichhoff) (Col.: Scolytidae), in India, trunk borer, *Anoplophora maculata* (Thomson) (Col.: Cerambycidae), in Taiwan, and fruit flies, *Bactrocera* spp. (Dip.: Tephritidae), in Asia.

5.9 Pests of Mangosteen (*Garcinia mangostana* L., Guittiferae)

The mangosteen originated in the Malay Peninsula and is cultivated mostly in Malaysia, Indonesia, Thailand, and the Philippines. It is one of the most sought after tropical fruits; relatively few pests attack it.

Common name: None
Scientific name: *Stictoptera cucullioides* Guenée (= *Steiria subobliqua, Steiria vaiablilis, Stictoptera timesia*)
Order: Lepidoptera

Family: Noctuidae
Distribution: South and Southeast Asia, Australia
Hosts: Mangosteen, *Clusia rosea, Mammea americana, Calophyllum* sp.
Life cycle: Eggs are laid individually on tender leaves. Mature larvae are about 30 mm in length with orange to reddish brown heads, and light to whitish green bodies with white dorsolateral and supraspiracular lines. Pupation takes place in the soil.
Damage: Larvae feed mostly on tender leaves. Infrequent defoliation occurs which usually does not require external control measures.
Natural enemies: No information is available.

Other major pests are:
Citrus leaf miner, *Phyllocnistis citrella* (See Pests of Citrus)
Coconut scale, *Aspidiotus destructor* (See Pests of Papaya)

Other pests of minor importance are: *Eupterote* sp. (Lep.: Eupterotidae) and *Orgyra* sp. (Lep.: Lymantridae) in Asia.

5.10 Pests of Passion Fruit (*Passiflora edulis* Sims, Passifloraceae)

Passiflora edulis is a vine species of passion flower that is native to Paraguay, Brazil, and northern Argentina. It is widely cultivated for its fruit in warmer frost-free areas throughout the world. The fruit is either yellow or dark purple at maturity with a soft, juicy interior filled with numerous seeds. The fruit is both eaten and juiced and is high in beta-carotene, vitamin C, potassium, and dietary fiber.

Common names: Red and black flat mite, False spider mite
Scientific name: *Brevipalpus phoenicis* (Geijskes)
Order: Acari
Family: Tenuipalpidae
Distribution: Asia, North Africa, Florida, Caribbean
Hosts: Citrus, guava, papaya, passion fruit, tea, coffee, and sweet potato
Life cycle: Eggs are reddish orange and are laid singly in cracks and crevices of the plant surface. Larvae hatch from eggs and development continues through the protonymph and deutonymph stages. Development time from egg to adult ranges from 10 to 27 days.
Damage: Feeding of this mite causes leaves to turn rusty brown and in severe infestations defoliation may occur.
Natural enemies: A number of predators have been reported to feed on these mites including the predatory mites, *Phytoseiulus macropilus, Amblyseius largoensis,* and *Mexecheles hawaiiensis,* and the beetle, *Sticholotis punctata.*

5.11 Pests of Pomegranate (*Punica granatum* L., Punicaceae)

Pomegranate is native to Persia, the Caucasus, and the Western Himalayas and is cultivated in Eastern and South Asia. India, Pakistan, Afghanistan, Iran, Egypt, Bangladesh, China, and Myanmar are major producers. It is made into juice in many countries, and the arils are eaten raw. The juice is rich in vitamin C, potassium, tannins, and flavonoids. Pomegranate butterfly and fruit piercing moths are its most serious pests.

Common names: Anar butterfly, pomegranate butterfly
Scientific name: *Deudorix (Virachola) isocrates* (Fabricius)
Order: Lepidoptera
Family: Lycaenidae
Distribution: India, adjoining countries
Hosts: Apple, *Ziziphus* sp., citrus, guava, litchi, loquat, peach, pear, plum, pomegranate, sapota, tamarind
Life cycle: This butterfly is bluish brown with an orange patch on the forewings and black spots on the hindwings. It lays single, shiny, oval eggs on the calyx of flowers and on small fruits. The eggs hatch in 7–10 days. Young larvae bore into the fruit and feed on seeds. Larvae are dirty brown and when fully grown they are 16–20 mm in length. Pupation occurs either inside the fruit or on the stalk. The life cycle varies from 1 to 2 months depending on the temperature.
Damage: The caterpillars cause damage by feeding inside the fruit. Holes made by the caterpillars provide entry points for bacteria and fungi which cause rotting. This pest may cause 40% or more damage.
Natural enemies: The larvae are parasitized by *Brachymeria euploeae* in India.

Common name: Bark borer
Scientific name: *Indarbela quadrinotata* (Walker)
Family: Metarbelidae
Order: Lepidoptera
Distribution: South and Southeast Asia
Hosts: Pomegranate, *Paraserianthes falcataria, Casuarina* spp., *Acacia* spp., *Albizia* spp., *Anacardium occidentale, Pongamia pinnata*
Life cycle: Eggs are deposited on the bark of trees and newly-hatched larvae bore into the bark. Larvae feed, cover their feeding sites with weblike material, and develop through eight larval stages before pupating.
Damage: Larvae feed on branches, causing the branches to wither and die. Larval tunnels and feeding scars caused by larval feeding serve as entry points for plant pathogens. Heavy infestations may cause significant damage.
Natural enemies: An entomogenous fungus, *Beauveria bassiana,* and a nematode, *Steinernema carpocapsae*, have been found attacking the bark borer.

Common name: Castor semilooper
Scientific name: *Achaea janata* (L.)
Order: Lepidoptera
Family: Noctuidae
Hosts: Pomegranate, castor, soybean, cotton, macadamia, litchi, tamarind, rose, croton, others
Distribution: Asia, Australia, New Zealand, Pacific Islands
Life cycle: The female moth lays green spherical eggs with deep striations singly on the leaves of host plants. Eggs hatch in 3–4 days. The larvae are semiloopers and have six instars. The larval stage lasts about 2 weeks. Pupation takes place in a cocoon spun in a folded leaf or under fallen leaves on the ground.
Damage: The adult moth is a secondary piercer of fruits. The caterpillars can defoliate plants.
Natural enemies: Eggs are parasitized by *Trichogramma minutum*, *Trichogramma evanescens*, and *Telenomus proditor*. Other parasitoids reported include: *Hyposoter exguae*, *Apanteles sundanus*, *Apanteles ruidus*, *Microplitis maculipennis*, *Microplitis similes*, *Euplectrus leucostomus*, *Paniscus ocellaris*, *Zamesochorus orientalis*, *Tetrastichus ophiusae*, *Rogas percurrens*, and *Enicospilus* sp. on larvae, and *Blondelia* (*Eucelatoria*) *armigera* and *Chaetogaedia monticola* on larvae and pupae.

Other major pests are:
Fruit piercing moth, *Eudocima phalonia* (See Pests of Citrus)
Nettle caterpillar, *Parasa philepida* (*lepida*) (See Pests of Beans)
Striped mealybug, *Ferrisia virgata* (See Pests of Tomato)

Other minor pests known to attack pomegranate: *Euproctis fraterna* (Moore) (Lep.: Lymantridae), *Bactrocera zonata* (Saunders) (Dip.: Tephritidae), *Myllocerus* spp. (Col.: Curculionidae)

5.12 Pests of Rambutan (*Nephelium lappaceum* L., Sapindaceae)

Rambutan is native to the Malaysian Archipelago. From there it has spread throughout South and Southeast Asia. It is grown in Malaysia, Thailand, India, Sri Lanka, Indonesia, and the Philippines. The fruits are consumed fresh or canned (Ooi *et al.*, 2002). The cocoa pod borer is one of the important pests of rambutan. Because this pest feeds on the fruit next to the fruit stalk, damage is usually not obvious.

Common name: Prominent moth
Scientific name: *Hyperaeschrella insulicola* (Kiriakoff)
Order: Lepidoptera
Family: Notodontidae
Distribution: Thailand, Malaysia
Hosts: Rambutan, cashew
Life cycle: Little is known about the biology of this pest. Eggs are laid on the leaves and caterpillars feed there. Pupation takes place on the ground in leaf litter (Fig. 5.12.1).

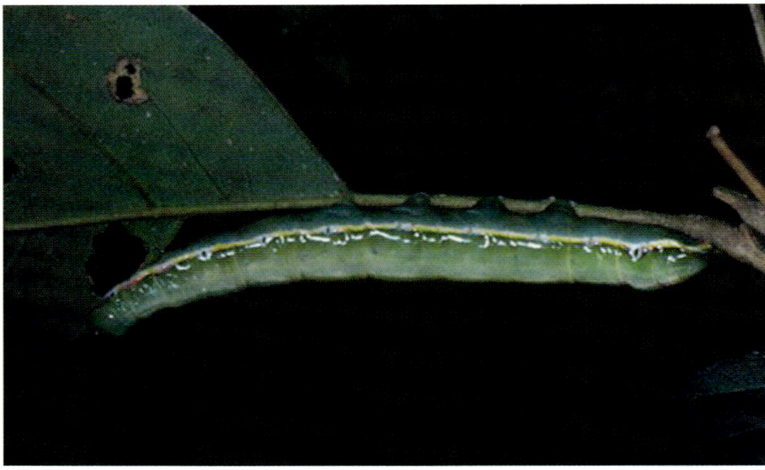

Fig. 5.12.1. Larva of the prominent moth, *Hyperaeschrella insulicola*, feeding on a rambutan leaf

Damage: Outbreaks of this pest occur in Malaysia, resulting in severe defoliation, especially during dry periods.
Natural enemies: No information is available.

Common name: Cocoa bark borer
Scientific name: *Squamura disciplaga* (Swinhoe) (*Indarbela disciplaga*)
Order: Lepidoptera
Family: Metarbelidae
Distribution: India, Malaysia, China
Hosts: Rambutan, pomegranate, cocoa, durian, soursop, rose apple, clove, mulberry
Life cycle: Eggs are laid singly on the bark of trees. Upon hatching, the young larvae feed on the bark and spin webs of frass over themselves. They feed at night and rest during the day. As the larvae grow, they bore into the stem where they pupate. The life cycle takes 3–4 months.
Damage: This is a serious pest of cocoa in Malaysia. Severe infestation can weaken the vigor of the trees.
Natural enemies: A braconid wasp was found to parasitize pupae in Malaysia.

Common name: Cocoa mealybug
Scientific name: *Cataenococcus hispidus* (Morrison) (*Pseudococcus hispidus*)
Order: Hemiptera
Family: Pseudococcidae
Distribution: Malaysia, Thailand
Hosts: Rambutan, durian, citrus, guava
Life cycle: Reproduction is both sexual and parthenogenetic. Females are ovoviviparous and pass through three nymphal instars, lasting 8, 6, and 7 days, respectively. The males pass through two nymphal, a prepupal, and a pupal instar of

7, 8, 2, and 4 days duration, respectively. The adult male is winged and lives for about 2 days while the female lives for about 23 days.
Damage: This is a minor pest. It does not cause serious economic damage.
Natural enemies: Little or no information is available.

Other major pests are:
Nettle caterpillar, *Parasa philepida* (*lepida*) (See Pests of Beans)
Atlas moth, *Attacus atlas* (See Pests of Guava)
Gelatine grub, *Chalcocelis albiguttatus* (See Pests of Durian)
Cocoa pod borer, *Conopomorpha cramerella* (See Pests of Cocoa)
Rose beetles, chafer beetles, *Adoretus compressus*, *Adoretus sinicus* (See Pests of Banana)
Yellow peach moth, *Conogetes punctiferalis* (*Dichocrocis punctiferalis*) (See Pests of Durian)

5.13 Pests of Santol (*Sandoricum koejapi* Merr. (*Sandoricum indicum* Cav.), Meliaceae)

Santol is a native of Southeast Asia and has spread to other tropical regions. It is commonly grown for local markets in the Philippines and Thailand. The blister mite is the most serious pest.

Common name: Santol blister mite
Scientific name: *Eriophyes sandorici* Nalepa
Order: Acarina
Family: Eriophyidae
Origin: Southeast Asia
Distribution: Philippines, Thailand
Life cycle: This mite infests the lower surface of young leaves. As the leaves grow, blister-like galls protrude from the upper surface. Mites remain in the aceria produced inside the gall. Even though the affected leaves do not fall prematurely, the normal function of the leaf is affected.
Damage: This pest was introduced to the Philippines sometime after 1950. Severe infestations cause poor fruit set.
Natural enemies: A fungus, *Hirsutella thompsonii*, is known to attack this eriophyid mite.

Scientific name: *Metanastria hyrtaca* Cramer
Family: Lasiocampidae
Order: Lepidoptera
Distribution: India, Nepal, Myanmar, Thailand, Malaysia, Indonesia, Vietnam, China, Taiwan
Hosts: Santol, *Syzygium cumini*, cashew, *Nyctanthes arbor-tristis*, *Gmelina arborea*, *Madhuca latifolia* (Aherkar *et al.*, 1997).

Life cycle: The adult female moth is brown with black markings on the wings. Eggs are laid in batches and each female lays about 330 eggs during her lifetime. Larvae pass through five to six instars ranging from 30 to 35 days. A thick layer of long gray hairs covers the body of the larvae. Fully grown larvae are about 65 mm long and 15 mm in diameter. Larvae feed on leaves at night and migrate to the trunk during the day. Pupation takes place on the tree trunk in a cocoon covered with larval hairs. Pupal duration is about 12 days and adults live up to 6 days.
Damage: Larvae feed on young leaves and often cause serious defoliation.
Natural enemies: An egg parasitoid, *Mesocomys orientalis*, was recorded in India.

Other major pests:
Atlas moth, *Attacus atlas* (See Pests of Guava)
Oriental fruit fly, *Bactrocera dorsalis* (See Pests of Mango)
Coffee mealybug, *Planococcus lilacinus* (See Pests of Soursop)

5.14 Pests of Sapodilla (*Manilkara zapota* (L.) P. van Royen, Sapotaceae)

Sapodilla is also known as chiku, chiko, and sapota. It is native to Southern Mexico, Central America, and the Caribbean. It is an evergreen tropical tree and resistant to wind. Large acreages of it are grown in India, Pakistan, Sri Lanka, and the Philippines. Fruits are eaten raw, and in Mexico the principal ingredient in chewing gum, or chicle, is extracted from the tree. Sapodilla is attacked by relatively few insect pests. Chiku bud borer and sapota leaf-webber may be considered regionally important.

Common name: Chiku bud and fruit borer
Scientific name: *Anarsia achrasella* Bradley
Family: Gelechiidae
Order: Lepidoptera
Distribution: India
Host: Sapodilla
Life cycle: The adult female lays eggs singly on the buds or on developing fruits. Each female lays 40–60 eggs in her lifetime. The duration of egg, larval, and pupal stages are 2–6, 11–15, and 6–8 days, respectively (Dhara and Tandon, 1994).
Damage: The borer feeds on buds and young fruits which can result in significant crop loss.
Natural enemies: A parasitoid, *Goniozus* (*Perisierola*) sp., was recorded in India.

Common names: Sapota leaf-webber, Chiku moth
Scientific name: *Nephopteryx eugraphella* Rogonot
Family: Pyralidae
Order: Lepidoptera
Distribution: India
Host: Sapodilla

Life cycle: Female moths lay eggs singly or in batches of 2–3 eggs on tender leaves or buds. A female may lay as many as 374 eggs in a week. Eggs hatch in 2–11 days. Larvae feed on leaves and buds that they web together. The larval stage lasts 13–60 days. Pupation takes place in the webbed leaves and lasts for 8 to 19 days. Seasonal temperatures determine the length of the life cycle (Nayar *et al.*, 1976; Atwal and Dhaliwal, 2009).
Damage: Larval feeding may cause serious damage to young terminals and buds, which reduces flowering and fruiting.
Natural enemies: The larval-pupal parasitoids, *Eurytoma* sp., *Xanthopimpla* sp., and *Cadurcia* sp., and predators such as *Calleida splendidula* and *Parena nigrolineata* were recorded in India (Sran and Sandhu, 1979).

Other major pests are:
Gold dust weevil, *Hypomeces squamosus* (See Pests of Mango)
Coffee mealybug, *Planococcus lilacinus* (See Pests of Soursop)
Anar butterfly, pomegranate butterfly, *Deudorix (Virachola) isocrates* (See Pests of Pomegranate)
Oriental fruit fly, *Bactrocera dorsalis* (See Pests of Mango)
Mango hoppers, *Idioscopus clypealis*, *Idioscopus niveosparsus*, *Idioscopus nitidulus*, *Idioscopus incertus* (See Pests of Mango)

5.15 Pests of Soursop (*Annona muricata* L., Annonaceae)

The soursop is native to Central America, the Caribbean, and northern South America. In some Southeast Asian countries it is grown on a commercial scale for producing juice. It is also eaten ripe, or used to make smoothies or ice cream. Although pests specific to this tree occur in its center of origin, they are not found in Asia. However, pests of other crops have adapted to feed on this fruit tree.

Common name: Coffee mealybug
Scientific name: *Planococcus lilacinus* (Cockerell)
Order: Hemiptera
Family: Pseudococcidae
Hosts: Cocoa, soursop, guava, citrus, rambutan, coffee, kapok, others
Distribution: India, Malaysia, Indonesia, Taiwan, the Philippines
Life cycle: Brownish red adults are oval to round. They bear live young and can reproduce without mating. A female deposits about 270 nymphs which are gregarious. They reach the adult stage in 37–50 days.
Damage: Infestation by this pest causes distortion of leaves and shortening of internodes. Honeydew produced causes sooty mold to develop.
Natural enemies: The coccinellid, *Scymnus* sp., and the parasitoid, *Anagyrus* sp., have been recorded in Malaysia. Different species of ants are known to attend this mealybug, of which *Oecophylla smaragdina* and *Dolichoderus thoracicus* are common.

Other major pests are:
Pink hibiscus mealybug, *Maconellicoccus hirsutus* (See Pests of Okra)
Cocoa bark borer, *Squamura disciplaga* (See Pests of Rambutan)
Mango webworms, *Orthaga exvinacea* (India) and *Orthaga icarusalis* (Malaysia) (See Pests of Mango)
Atlas moth, *Attacus atlas* (See Pests of Guava)
Black citrus aphid, *Toxoptera aurantii* (See Pests of Citrus)
Striped mealybug, *Ferrisia virgata* (See Pests of Tomato)
Fruit piercing moth, *Eudocima phalonia* (See Pests of Citrus)
Citrus mealybug, *Planococcus citri* (See Pests of Citrus)
Cotton aphid, melon aphid, *Aphis gossypii* (See Pests of Eggplant)
Spiraling whitefly, *Aleurodicus dispersus* (See Pests of Guava)
Tailed jay, *Graphium agamemnon* (See Pests of Sweetsop)
Tea mosquito, *Helopeltis theivora* (See Pests of Tea)

Other pests of regional importance are: Tree hoppers, *Tricentrus capreolus* (Walker), and *Orthobelus flavipes* Uhler (Hem.: Membracidae)

5.16 Pests of Sweetsop, Atis, or Custard Apple (*Annona squamosa* L., Annonaceae)

Annona squamosa is a semi-evergreen shrub or small tree that is native to the tropical Americas, but is more widely grown in India, Pakistan, and the Philippines. It is the most widely cultivated of three species of *Annona*, the other two being *Annona reticulata* and *Annona cherimola* (whose fruits often share the same name). The unusual heart-shaped fruit has a surface that is divided into small, knobby scales that can be broken apart easily when the fruit is ripe, exposing the creamy, sweet, custardlike pulp which encloses the small black seeds.

Common name: Atis fruit borer
Scientific name: *Anonaepestis* (*Heterographis*) *bengalella* Ragonot
Order: Lepidoptera
Family: Pyralidae
Hosts: Sweetsop
Distribution: Philippines, India, Indonesia
Life cycle: Eggs are laid singly on the immature fruit. Upon hatching, the caterpillars bore into the fruit. Fully developed whitish gray caterpillars are about 20 mm long. Pupation takes place in the larval tunnel (Cendaña *et al.*, 1984).
Damage: This is a serious pest of atis in the Philippines. Young infested fruits die and parts of older infested fruit dry up. In some seasons, as much as 50–90% of the fruit is damaged.
Natural enemies: None reported.

Common name: Tailed jay
Scientific name: *Graphium agamemnon* L.
Order: Lepidoptera
Family: Papilionidae
Hosts: Cinnamon, *Annona* spp., *Michelia* spp.
Distribution: South and Southeast Asia, Australia
Life cycle: The adult is a brownish black butterfly with many green spots on the wings. Eggs are laid singly on tender leaves. Young larvae feed on tender leaves, and start feeding on older leaves as they grow. Early stages of the larvae are black and cream, resembling bird droppings, and later stages are velvety green-brown, matching new shoots. The larval stage lasts 10–20 days. The pupa is a chrysalis attached to a leaf or twig. Pupal duration is 1–2 weeks (Fig. 5.16.1).
Damage: This butterfly is a minor pest and causes minimal damage to its host plants.
Natural enemies: Eggs are parasitized by *Trichogramma* sp., larvae by *Apanteles papilionis*, and pupae by *Pteromalus* sp. Predators include the wasp *Polistes herebreus* and the preying mantis *Creobroter gemmatus* (Nayar *et al.*, 1976; Kalshoven, 1981).

Other major pests are:
Coconut scale, *Aspidiotus destructor* (See Pests of Papaya)
Striped mealybug, *Ferrisia virgata* (See Pests of Tomato)
Coffee mealybug, *Planococcus lilacinus* (See Pests of Soursop)
Tea mosquito, *Helopeltis theivora* (See Pests of Tea)
Atlas moth, *Attacus atlas* (See Pests of Guava)

Fig. 5.16.1. Caterpillar of tailed jay, *Graphium agamemnon*

5.17 Pests of Tamarind (*Tamarindus indica* L., Fabaceae)

The tamarind is indigenous to the drier savannas of tropical Africa and is economically important in Southeast Asia. It was naturalized long ago in tropical Asia and is now cultivated in all tropical countries. It is widely grown in Asia as an edible fruit and condiment. Tamarind is attacked primarily by polyphagous insect pests such as lymantriid caterpillars and grasshoppers.

Common name: Tussock moths
Scientific names: *Dasychira inclusa* Walker, *Dasychira mendosa* Hübner
Order: Lepidoptera
Family: Lymantriidae
Distribution: South and Southeast Asia.
Hosts: Polyphagous. Tamarind, citrus and many others
Life cycle: Eggs are laid in large numbers on leaves where emerging caterpillars feed. Caterpillars are recognized by tufts of urticating hairs on the dorsal part of the body (Shepard *et al.*, 1999) and may grow up to 40–50 mm for *D. inclusa* and 30–40 mm for *D. mendosa*. The larval stage of *D. mendosa* lasts 21–28 days and the pupation period lasts 11–12 days (Fig.5.17.1).
Natural enemies: Yunus and Ho (1980) recorded *Brachymeria euploeae* and *Carcelia buitenzorgiensis* from *D. inclusa*. *Apanteles mendosae* was recorded attacking *D. mendosa* in Java and Malaysia.

Other major pests are:
Coffee mealybug, *Planococcus lilacinus* (See Pests of Soursop)
Anar butterfly, pomegranate butterfly, *Deudorix (Virachola) isocrates* (See Pests of Pomegranate)

Fig. 5.17.1. Larva of *Dasychira mendosa*

Yellow peach moth, *Conogetes punctiferalis* (*Dichocrocis punctiferalis*) (See Pests of Durian)
Black citrus aphid, *Toxoptera aurantii* (See Pests of Citrus)
Oriental migratory locust and grasshoppers, *Locusta migratoria, Atractomorpha* sp. (See Pests of Amaranths)
Javanese grasshopper or Malaysian locust, *Valanga nigrocornis* (See Pests of Avocado)

Other pests of regional importance are: Oriental scale, *Aonidiella orientalis* (Newstead) (Hem.: Diaspididae), and Black scale, *Saissetia oleae* Olivier (Hem.: Coccidae)

5.18 Pests of Ziziphus (*Ziziphus jujuba* Miller, Rhamnaceae)

Ziziphus jujuba, commonly called Chinese date or Indian date, is a small deciduous thorny tree in the buckthorn family. It is thought to have originated in southern Asia and is used primarily as a fruiting shade tree. The fruit is a small oval green drupe that turns brown when ripe. Freshly harvested fruit as well as candied dry fruit is eaten as a snack.

Common name: Ber fruit fly
Scientific name: *Carpomyia vesuviana* Costa
Order: Diptera
Family: Tephritidae
Distribution: India, Pakistan, southern Italy
Hosts: This fly is most destructive against *Ziziphus mauritiana* and *Z. jujuba* in India and *Ziziphus sativa* in Italy.
Life cycle: This pest is active during winter months and remains inactive in the soil as pupae from April to August. Flies emerge from pupae from August to mid-November. The flies make cavities in the skin of fruit and deposit one or two spindle shaped creamy white eggs 1 mm below the skin. Eggs hatch in 2–3 days and the maggots feed on the flesh of the fruit, making galleries toward the center. The larvae are full-grown in 7–10 days and emerge from the fruit by cutting holes in the skin. They drop to the soil and pupate 6–15 cm beneath the soil surface. The pupal stage lasts 14–30 days. The shortest life cycle, from egg to emergence of the adult, is 24 days. There are 2–3 broods in a year.
Damage: Damage is caused by the larvae, which are creamy white and slightly smaller than most other fruit flies. Fruits will rot near the stone where the larvae feed. Fly populations build up during the winter, and in February and March late maturing fruit can be entirely destroyed.

Common name: Ber beetles
Scientific names: *Adoretus pallens* Arrow and *Adoretus nitidus* Arrow
Order: Coleoptera
Family: Scarabaeidae
Distribution: Northern India, Pakistan

Hosts: These beetles are polyphagous, but prefer the ber tree (*Ziziphus jujuba*) and grapes.

Life cycle: Beetles are active during the summer and pass the winter in the larval stage. The adults appear in April and May and lay eggs singly in soil near the host plants. Eggs hatch in 6–9 days. The white grubs feed on roots of grasses and other vegetable matter in the soil. When full-grown, the grubs make an earthen cell and hibernate through the whole winter. Pupation takes place during the next April and the adults emerge after 11–12 days. There is only one generation per year.

Damage: The beetles cut round holes in the leaves. Ber trees are sometimes so heavily attacked that all the foliage is lost and these trees do not bear any fruit.

6 Pests of Other Crops

6.1 Pests of Cashew (*Anacardium occidentale* L., Anacardiaceae)

The cashew is native to Brazil. In the mid 16th century, the Portuguese took it to Goa, India, from where it has spread to the Old World tropics (Yaacob and Subhadrabandhu, 1995). Vietnam, India, and Indonesia are the major producers of cashew in Asia. Juice made of cashew apple constitutes a minor industry in some countries; however, cashew nut is the major international trade item. The nut contains both saturated and unsaturated fat. It is a popular snack and is also used in many cuisines. Major pests of this crop are stem borer, tea mosquito, and bark borer.

Common name: Cashew stem and root borer
Scientific name: *Plocaederus ferrugineus* L.
Order: Coleoptera
Family: Cerambycidae
Distribution: Asia, Africa
Hosts: Cashew, *Spondias mombin*
Life cycle: The adult beetle lays eggs under the loose bark of the trunk. Newly emerged larvae bore into the bark and as they mature they bore into the wood. Tunnels are plugged with chewed fiber, excreta, and gum exudates and extrusion of fibrous frass can be seen from damaged areas. Fully grown larvae measure about 7.5 cm and they usually tunnel downwards to the root zone. Pupation is in a calcareous cocoon buried in the frass and soft wood tissue of the trunk or roots (Asogwa *et al.*, 2009).
Damage: Tunneling of the trunk weakens the tree. Heavy infestations lead to the death of mature trees.
Natural enemies: The fungal pathogens, *Beauveria bassiana*, and *Metarhizium anisopliae*, will infect this pest and have been used successfully in field trials.

Common name: Bark borer
Scientific name: *Indarbela tetraonis* M.
Order: Lepidoptera
Family: Metabelidae
Distribution: India, Myanmar, Bangladesh, Sri Lanka
Hosts: Polyphagous including mango, moringa, guava, citrus, pomegranate, litchi
Life cycle: Eggs are laid in clusters under the loose bark of the trees. Eggs hatch in 8–10 days. Newly emerged larvae bore into the bark and mature larvae tunnel in and feed on bark, and also bore into the wood. The exposed part of the winding tunnel made by the mature larva is covered with webs holding chewed bark and excreta. The larval stage lasts 9–11 months. Pupation takes place in the tunnel and lasts 3–4 weeks. The adult emerges, leaving a portion of the pupal case protruding from the bore hole.
Damage: Larval boring of the bark causes severe interruption of the sap flow. Severe infestations cause branches and, in some cases, trees to die.
Natural enemies: A dipteran parasitoid, *Zenillia heterusiae,* which attacks larvae has been reported from Sri Lanka.

Other major pests are:
Tea mosquito, *Helopeltis antonii* (See Pests of Tea)
Chili thrips, Hevea thrips, *Scirtothrips dorsalis* (See Pests of Pepper)
Red-banded thrips, *Selenothrips rubrocinctus* (See Pests of Mango)
Mango trunk and branch borer, *Rhytidodera simulans* (See Pests of Mango)
Mango tip moth, *Penicillaria jocosatrix* (See Pests of Mango)
Prominent moth, *Hyperaeschrella insulicola* (See Pests of Rambutan)
The baron, *Euthalia aconthea* (See Pests of Mango)
Mango webworms, *Orthaga exvinacea* (India), *Orthaga icarusalis* (Malaysia) (See Pests of Mango)

6.2 Pests of Cocoa (*Theobroma cacao* L., Sterculiaceae)

Cocoa is South American in origin, but 70% of the current world cocoa production is from West Africa. This crop is mostly grown between the Tropic of Cancer and Tropic of Capricorn. Indonesia is the major producer of cocoa in Asia, followed by Papua New Guinea, Malaysia, India, the Philippines, and the Solomon Islands. Cocoa is grown both on a plantation scale and in smallholder agriculture in Asia. Cocoa pod borer and tea mosquito are the major pests of cocoa in Asia.

Common name: Cocoa pod borer
Scientific name: *Conopomorpha cramerella* (Snellen)
Order: Lepidoptera
Family: Gracillariidae
Distribution: This pest occurs in almost all countries in South and Southeast Asia where hosts are available.

Hosts: Cocoa and also fruits of *Pometia* sp. (Sapindaceae), *Nephelium* spp. (Sapindaceae), *Cynometra* sp. (Leguminosae), and *Cola* sp. (Sterculiaceae). Of these, only plants within Sapindaceae are hosts in Southeast Asia.

Life cycle: The flat, oval, 0.5 mm eggs are deposited by female moths onto the surface of pods. The larva hatches in about 7 days and immediately bore into the pod after tunneling through the floor of the egg. There are 4–6 instars which last 14–18 days when they develop in cocoa. Fully mature, 12 mm long larvae exit the pod and pupate on the pod, in green or dried leaves, or in other debris. The pupal stage lasts up to 8 days. Moths are nocturnal and rest during the day under leaves or branches (Figs 6.2.1, 6.2.2, 6.2.3).

Damage: The cocoa pod borer is a key pest of cocoa and a major constraint to cocoa production in Southeast Asia. Damage is caused by the tunneling larvae that cause undersized, poor quality beans. In addition, the beans stick together and are hard to remove from the pulp (Fig. 6.2.4).

Natural enemies: More than 24 different parasitoid species have been recorded from this pest from Malaysia, the Philippines, and Sri Lanka (Ooi, 1988). Most were pupal parasitoids from the family Ichneumonidae; these include *Diaglytidea roepkei*, *Photoptera erythronota*, *Goryphus javanicus*, *Goryphus fasciatipennis*, *Goryphus* sp.,

Fig. 6.2.1. Adult of cocoa pod borer *Conopomorpha cramerella*

Fig. 6.2.2. Larva of cocoa pod borer

Fig. 6.2.3. Pupa of cocoa pod borer

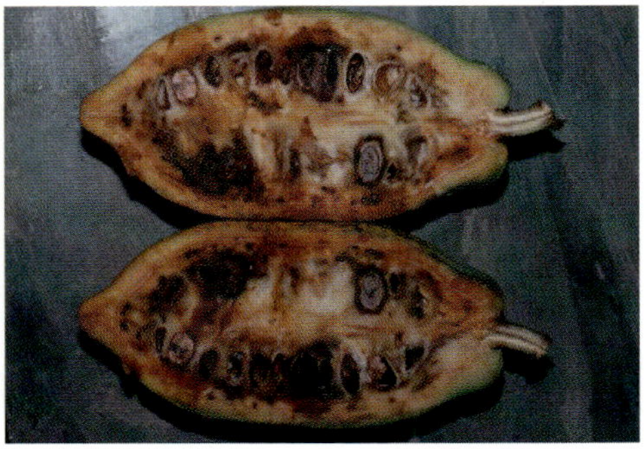

Fig. 6.2.4. Pod borer damaged cocoa fruit

Paraphylax fasciatipennis, three other *Paraphylax* spp. and *Xanthopimpla* sp. Other parasitoids were from the families Trichogrammatidae, Braconidae, Ceraphronidae, Chalcidae, Encyrtidae, Eulophidae, Elasmidae, and Phoridae. Ooi (1988) has provided a list of the species of egg, larval, and pupal parasitoids. At least one parasitoid, *Trichogrammatoidea bactrae fumata*, has been studied for its possible use in a biological control program. There is a significant complex of generalist predators that feed on the eggs and pupae. These include *Oecophylla smaragdina, Iridomyrmex anceps, Anoplolepis longipes, Crematogaster* sp., and *Dolichoderus thoracicus*. Several fungi have been found to infect cocoa pod borer larvae and pupae. The most effective is *Beauveria bassiana*.

Common names: Cocoa mirid or capsid, mosquito bug, tea mosquito
Scientific names: *Helopeltis theivora* (*Helopeltis theobrome*) Waterhouse and *Helopeltis schoutedeni* Reuter
Order: Hemiptera
Family: Miridae
Distribution: South and Southeast Asia and Africa
Hosts: Cocoa, mango, cashew, guava, tea, *Acalypha* spp., Japanese cherry, neem, and many others.
Life cycle: The 5–7 mm long adult mirids deposit eggs just below the surface of the pod or into tender stems at the rate of about 4–10 eggs per day and a female will lay about 200 eggs during her lifetime. Eggs hatch in about 7 days, and nymphs pass through 5 instars in 18 days. Populations peak during the rainy season (Sundararaju and Sundara Babu, 1999; Hazarika *et al.*, 2008).
Damage: Nymphs and adults feed similarly by sucking sap from the plant. In guava, flower buds and fruits are attacked. Insect density and cocoa crop loss relationships were studied in Malaysia. Very young fruit (cherelles) were most seriously affected; one feeding by the mirid caused these fruit to abort. Even if young pods survive after feeding by the mirid, they were smaller and contain smaller seeds. More mature pods are not as much affected by mirid feeding based on results from caging the insects onto the cocoa pods. Damage to pods at any stage may provide a route for invasion by plant pathogens. In addition, these mirids feed on the plants' soft growing tips and leaf petioles, which may result in dieback and canopy degeneration. However, the major damage is to young developing pods.
Natural enemies: *Beauveria bassiana* has been isolated from this insect in Malaysia and laboratory studies showed that it grew well on adult bugs. In plots where the black cocoa ant, *Dolichoderus thoracicus*, was removed by insecticides, damage by the cocoa mirids was significantly higher, which indicated the importance of these ants as predators. Other studies have shown that the weaver ant, *Oecophylla smaragdina*, can provide almost complete control of *H. theivora*. Some of the predators reported from India were *Chrysoperla carnea, Oxyopes* sp., *Plexippus* sp., and *Marpissa* sp.

Other major pests are:
Coffee borer, *Zeuzera coffeae* (See Pests of Coffee)
Cocoa bark borer, *Squomura disciplaga* (See Pests of Rambutan)
White moth cicada, *Lawana conspersa* (See Pests of Citrus)
Cluster caterpillar or army worm, *Spodoptera litura* (See Pests of Tomato)

Other pests of regional importance: *Darna diducta* (Snellen), *Darna trima* (Moore), *Parasa philepida* (Cramer), *Setora nitens* Walker (Lep.: Limacodidae), *Conogetes* (*Dichocrocis*) *punctiferalis* (Guenee) (Lep.: Pyralidae), *Pagodiella hekmeyeri* Heylaerts (Lep.: Psychidae), *Rastrococcus spinosus* (Robinson) (Hem.: Pseudococcidae), and *Valanga nigricornis* (Burmeister) (Orthoptera: Acrididae)

6.3 Pests of Coffee (*Coffea arabica* L. and *Coffea canephora* Pierre, Rubiaceae)

Coffea arabica originated in the highlands of Ethiopia, southeastern Sudan, and northern Kenya. *Coffea canephora* (robusta) originated in the equatorial lowland forests of Africa. Arabica coffee grows better at 1,000–2,000 meters and robusta thrives below 700 meters in the tropics. Currently, coffee is planted on over 10 million hectares in 50 or more countries (Vega, 2008). Arabica coffee is superior in quality and accounts for 75% of the world's production. Coffee is one of the most traded commodities in the world, and it is an important export crop of many developing countries. The major pest of coffee is the coffee berry borer. Coffee stem borer is important in some regions in Asia.

Common name: Coffee berry borer
Scientific name: *Hypothenemus hampei* (Ferrari)
Order: Coleoptera
Family: Scolytidae
Host: Coffee
Distribution: Africa, Central and South America, India, Indonesia, Malaysia, the Philippines, Thailand, Laos, Vietnam, Cambodia, Sri Lanka, Tahiti, New Caledonia, Phonpei, Saipan, Kona (Hawaii)
Life cycle: The adult female cuts a hole in the tip of the berry as it begins to ripen and then tunnels into and lays eggs inside the berry. Duration of egg, larval, and pupal stages are 5–9, 10–26, and 4–9 days, respectively. The developmental period from egg-laying to adult is about 25–35 days (Waterhouse, 1998) (Figs 6.3.1, 6.3.2, 6.3.3).
Damage: Considerable damage is caused when young berries are bored by the adults, as they fall from the plant. Larvae develop in seeds and these infested seeds either do not pass through the curing and grading process or, if they do, they will go to the lowest grade. Crop losses of 40 to 90% have been reported in plantations where control measures are not adopted (Le Pelley, 1968).
Natural enemies: Important parasitoids are *Prorops nasuta*, *Cephalonomia stephanoderis*, *Phymastichus coffea*, and *Heterospilus coffeicola*. *Prorops nasuta* was introduced to Java in 1923, and Sri Lanka in 1938, but it did not establish (Le Pelley, 1968). It was also introduced to Latin America, the Caribbean, Oceania, and Asia with limited success (Vega *et al.*, 2009). *Cephalonomia stephanoderis* was introduced from West Africa to more than 20 countries (Vega *et al.*, 2009). *Phymastichus coffea*, collected in Africa, was introduced to many countries in Latin America and to India.

Fig. 6.3.1. Coffee berry borer, *Hypothenemus hampei*

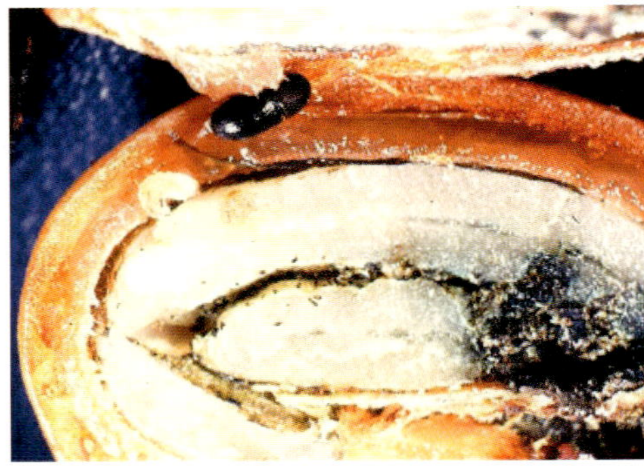

Fig. 6.3.2. Coffee berry borer adult and larva in a berry

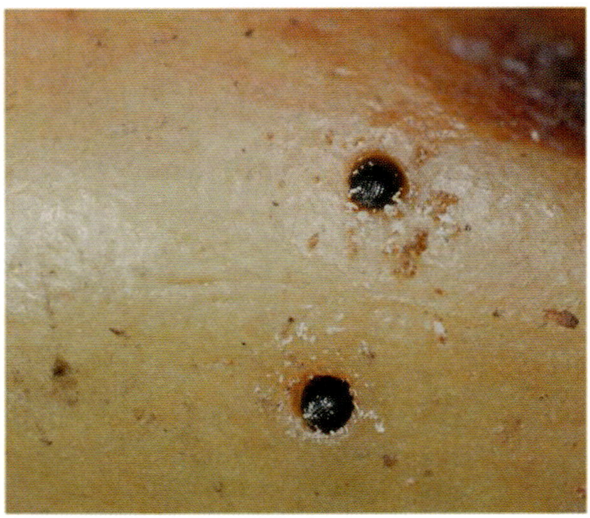

Fig. 6.3.3. Coffee berry with borer holes

Common name: Coffee stem borer
Scientific name: *Xylotrechus quadripes* Chevrolat
Order: Coleoptera
Family: Cerambycidae
Distribution: India, Nepal, Southeast Asia
Hosts: Coffee, teak, *Olea dioica*
Life cycle: Eggs are elongate, oval, 1.25 mm in length and are laid in groups of one to ten in cracks and crevices in the bark. They are milk white when laid and later turn slightly yellow. The egg stage lasts 9 to 15 days. The grub on hatching is white, and later turns yellow. The grub bores into the bark for about 2 months and then into the wood. The larval stage lasts for about 9 months and the pupal stage for 3 weeks to a

month. After emergence, adults remain in the pupal chamber for 3 to 7 days. Adults are active fliers.
Damage: The tunneling of the grubs results in the death of young plants and weakening of older plants.
Natural enemies: Parasitoids reported include *Metapelma* sp. in India, *Epixorides caerulescens* in Sri Lanka, and *Mysepyris grandiceps, Sclerodermus domesticus, Doryctes bistriatus, Doryctes brevipetiolus, Doryctes strioliger, Doryctes picticeps, Doryctes tristriatus, Pristodoryctes striativentris, Promiscolus sesquistriatus, Eurytoma xylotrechi, Pristaulacus nigripes* var. *duporti*, and *Paraglypta tubigera* in Southeast Asia. Le Pelley (1968) recommends a study of the full distribution of the borer and of its parasitoids throughout its range with the objective of introducing parasitoids to some countries such as India where only one parasitoid has been reported.

Common names: Clearwing moth, Bee hawkmoth
Scientific name: *Cephanodes hylas* L.
Order: Lepidoptera
Family: Sphingidae
Distribution: Asia and Africa
Hosts: Coffee, *Gardenia* sp.
Life cycle: The adult female lays single spherical eggs on young leaves and is capable of laying about 90 eggs. The egg, caterpillar, and pupal stages last for 3, 21, and 13 days, respectively. Pupation takes place in the soil or in the debris beneath the plant. The pupal stage lasts for about 13 days (Fig. 6.3.4).
Damage: This pest causes defoliation of coffee bushes when there is an outbreak.
Natural enemies: Eggs are parasitized by *Ooencyrtus malayensis* and the caterpillar by *Exorista sorbillans* in Malaysia. The reduviid, *Sycanus leucomesus*, has been noted to feed on the caterpillar (Le Pelley, 1968).

Fig. 6.3.4. Clearwing moth, *Cephanodes hylas*

Common name: Coffee borer
Scientific name: *Zeuzera coffeae* Nietner
Order: Lepidoptera
Family: Cossidae
Distribution: Asia
Hosts: Coffee, tea, loquat, cotton, orange, guava, teak, and others
Life cycle: Eggs are laid in masses and an adult is capable of laying over 1,000 eggs. The egg stage lasts 10–11 days. Young larvae first bore into green young shoots and as they grow migrate to thicker branches. The caterpillar stage lasts from 81 to 151 days (Le Pelley, 1968) (Figs 6.3.5, 6.3.6, 6.3.7).
Damage: This caterpillar is an important branch and stem borer of coffee and cocoa. The boring in the stems results in desiccation of shoots and eventual dying. This insect was considered to be a serious pest of coffee in India in the early 1900s; however, it is no longer considered a serious pest.
Natural enemies: The parasitoids, *Amyosoma leuzerae* from Indonesia and India and *Bracon* sp., *Iphiaulax* sp., and *Glyptomorpha* sp. from Malaysia, have been recorded.

Fig. 6.3.5. *Zeuzera coffeae* moth

Fig. 6.3.6. Larva of *Zeuzera coffeae*

Fig. 6.3.7. Pupa of *Zeuzera coffeae* in the stem

Other major pests are:
Fruit flies, *Bactrocera* spp. (See Pests of Mango)
White moth cicada, *Lawana conspersa* (See Pests of Citrus)
Pink hibiscus mealybug, *Maconellicoccus hirsutus* (See Pests of Okra)
Citrus mealybug, *Planococcus citri* (See Pests of Citrus)
Rose beetles or chafer beetles, *Adoretus compressus* and *Adoretus sinicus* (See Pests of Banana)
Nettle caterpillar, *Parasa philepida* (*lepida*) (See Pests of Beans)
Black citrus aphid, *Toxoptera aurantii* (See Pests of Citrus)

6.4 Pests of Tea (*Camellia sinensis* L., Theaceae)

Tea is a native of South and Southeast Asia and originated in the region that joins northeastern India, north Burma, southwest China, and Tibet. It contains flavonoids, amino acids, vitamins (C, E, and K), caffeine, and polysaccharides. China, India, and Sri Lanka are the major tea producers in Asia. Tea is grown in the hills up to 1,500 meters in the tropics. At these altitudes the plant grows more slowly and acquires a better flavor. The important pests of tea are tea mosquitoes and mites.

Common names: Scolytid beetle, shot hole borer
Scientific name: *Xyleborus fornicatus* Eichhoff
Order: Coleoptera
Family: Scolytidae
Distribution: South and Southeast Asia
Life cycle: This borer is a shiny dark brown to black, 2 mm long beetle and is known to attack several species of plants. It is considered to be a serious pest of tea in Sri Lanka. Mated females bore into the branches and inoculate the fungus, *Monacrosporium ambrosium*. When the bore hole reaches the center of the branch, the beetle forms an irregular breeding chamber and fills it with eggs. It will continue to expand the chamber and fill it with progeny which can be found at different stages of development.
Damage: Construction of galleries causes direct damage to the branches and also allows accumulation of wood rot which debilitates the plant.
Natural enemies: No information is available.

Common name: Tea red spider mite
Scientific name: *Oligonychus coffeae* (Nietner)
Order: Acarina
Family: Tetranychidae
Distribution: Asia, Australia, United States
Hosts: Polyphagous
Life cycle: The adult female lays an average of 90 eggs at the rate of 4–6 per day. Eggs are bright red and 0.11 mm in diameter. The developmental stages include a six-legged larva, a protonymph, and a deutonymph. The adult female is elliptical in shape, crimson in colour, and 0.3 to 0.4 mm in length. The male is smaller, 0.25–0.3

mm in length, and the abdomen is tapered to a point at the posterior end. Overlapping generations are found on the host leaves (Cranham, 1966a).
Damage: Affected leaves show bronze coloration on the upper side and close examination reveals webbing, red mites, and white cast skins.
Natural enemies: The predators, *Stethorus gilviforn*, *Micromus timidus*, and *Chrysoperla carnea* have been reported to feed on these mites (Das *et al.*, 2010).

Common name: Purple mite
Scientific name: *Calacarus carinatus* (Green)
Order: Acarina
Family: Eriophyidae
Distribution: India, Sri Lanka, Indonesia, Indochina
Life cycle: This mite is mostly found on older leaves and on both surfaces of leaves. The adult female mite is deep purple in color with five white waxy longitudinal ridges on the abdomen. The body is elongate and spindle shaped, and slightly broader at the anterior end. Adult females have two pairs of legs. Adult males are much shorter than females. Eggs are flat and there are three developmental stages. The life cycle is completed in 6–13 days depending on the temperature.
Damage: Affected leaves show a purplish bronze discoloration and white cast skins.
Natural enemies: None reported.

Other major pests are:
Tea mosquito, *Helopeltis theivora* (See Pests of Cocoa)
Red and black flat mite, *Brevipalpus phoenicis* (See Pests of Passion Fruit)
Yellow tea mite, broad mite, *Polyphagotarsonemus latus* (See Pests of Pepper)
Rose beetles, chafer beetles, *Adoretus compressus*, *Adoretus sinicus* (See Pests of Banana)
White moth cicada, *Lawana conspersa* (See Pests of Citrus)
Nettle caterpillar, *Parasa philepida* (*lepida*) (See Pests of Beans)
Tea tortrix, coffee leaf roller, *Homona coffearia* (See Pests of Citrus)
Brown citrus aphid and Black citrus aphid, *Toxoptera citricidus* and *Toxoptera aurantii* (See Pests of Citrus)

Other pests of minor importance are: Looper caterpillars, *Buzura suppressaria* Guen., *Hyposidra talaca* (Walker), and *Hyposidra infixaria* Walker (Lep.: Geometridae), Red slug caterpillar, *Eterusia magnifica* Butl. (Lep.: Limacodidae), Flush worm, *Cydia leucostoma* Meyrick (Lepidoptera: Tortricidae), Leaf roller, *Caloptelia theivora* Walsingham (Lepidoptera: Gracillariidae), Leafhopper (*Empoasca flavescens* Fabr.) (Hem.: Cicadellidae), and thrips (*Mycterothrips setiventris* (Bagnall) and *Scirtothrips dorsalis* Hood) (Thysanoptera: Thripidae).

References

Aherkar, S.K., Sarode, S.V., and Katole, S.R. (1997) Biology of *Metanastria hyrtaca* Cram. (Lepidoptera: Lasiocampidae) on jamun, *Syzygium cumini*. *Journal of Entomological Research* 21, 29–34.

Alam, S.N., Dutta, N.K., Ziaur Rahman, A.K.M., and Sarker M.A. (2006) *Annual Report, 2005–2006*. Division of Entomology. BARI, Joydebpur, Gazipur, Bangladesh, 86 pp.

AQIS. (2002) *Northern Australian Quarantine Strategy, Quarantine Pocket Guide, Essential Information for Residents of Northern Australia*. Australian Quarantine and Inspection Service, Department of Agriculture, Fisheries and Forestry, Australia, 37 pp.

Asogwa, E., Anikwe, J.C., Ndubuaku, T.C.N., Okelana, F.A., and Hammed, A. (2009) Host plant range and morphometerics descriptions of an emerging insect pest of cashew, *Plocaederus ferrugineus* L. (Coleoptera: Cerambycidae) in Nigeria: A preliminary report. *International Journal of Sustainable Crop Production* 4, 27–32.

Atwal, A.S. and Dhaliwal, G.S. (2009) *Agricultural Pests of South Asia and Their Management*. Kalyani Publishers, New Delhi, 616 pp.

Awadallah, K.T., Ammar, E.D., Tawfik, M.F.S., and Amirat, R. (1979) Life-history of the white mealy-bug *Ferrisia virgata* (Ckll) (Homoptera, Pseudococcidae). *Deutsche Entomologische Zeitschrift* 26, 101–110.

Bosco, D. and Caciagli, P. (1998) Bionomics and ecology of *Bemisia tabaci* (Sternorrhyncha: Aleyrodidae) in Italy. *European Journal of Entomology* 95, 519–527.

Braithwaite, B.M. (1958) Ground spray treatments for control of the banana beetle borer (*Cosmopolites sordidus* Germ.). *Journal of the Australian Institute of Agricultural Science* 24, 27–34.

Callan, E.McC. (1943) Natural enemies of the cacao-thrips. *Bulletin of Entomological Research* 34, 313–321.

Capinera, J.L. (2001) *Handbook of Vegetable Pests*. Academic Press, San Diego, 729 pp.

Cendaña, S.M., Gabriel, B.P., and Magallona, E.D. (1984) *Insect Pests of Fruit Plants in the Philippines*. College of Agriculture, University of the Philippines at Los Baños, Philippines, 86 pp.

Cho, C.M., Huang, C.J., Lo, C.F., Kou, G.H., and Wang, C.H. (1996) Characterization of *Perina nuda* nucleopolyhedrovirus (PenuNPV) polyhedrin gene. *Journal of Invertebrate Pathology* 67, 259–266.

Cranham, J.E. (1966a) Tea pests and their control. *Annual Review of Entomology* 11, 491–514.

Cranham, J.E. (1966b) *Insect and Mite Pests of Tea in Ceylon and Their Control*. Monographs on Tea Production in Ceylon Vol. 6, 122 pp.

Das, S., Roy, S., and Mukhopadhyay, A. (2010) Diversity of arthropod natural enemies in the tea plantations of North Bengal with emphasis on their association with tea pests. *Current Science* 99, 1457–1463.

De Barro, P.J. (1995) *Bemisia tabaci Biotype B: A Review of its Biology, Distribution and Control.* Commonwealth Scientific and Industrial Reseach Organization, Canberra, Australia, Technical Paper No. 33, 57 pp.

Dhara, J.B. and Tandon, P.L. (1994) Bionomics of sapota bud borer, *Anarsia achrasella* (Bradley) (Lepidoptera: Gelechiidae). *Journal of Entomological Research* 18, 135–138.

Eamrucksa, N. (1987) *Mass Rearing Technique, Biology and Crop Loss Assessment in Mungbean Damaged by Coreid Bugs,* Cletus trigonus *Thunb. (Hemiptera: Coreidae).* Kasetsart University Graduate School, Bangkok, Thailand, 71 pp.

Esguerra, N.M. and Gabriel, B.P. (1969) *Insect Pests of Vegetables*. College of Agriculture, University of the Philippines, Los Baños, 106 pp.

Follett, P.A. and Gabbard, Z. (2000) Effect of mango weevil (Coleoptera: Curculionidae) damage on mango seed variability in Hawaii. *Journal of Economic Entomology* 93, 1237–1240.

Foottit, R.G., Maw, H.E.L., Pike, K.S., and Miller, R.H. (2010) The identity of *Pentalonia nigronervosa* Coquerel and *P. caladii* van der Goot (Hemiptera: Aphididae) based on molecular and morphometric analysis. *Zootaxa* 2358, 25–38.

George, R.A.T. (2011) *Tropical Vegetable Production*. CAB International, Wallingford, UK, 225 pp.

Gerling, D., Alomar, O., and Arno, J. (2001) Biological control of *Bemisia tabaci* using predators and parasitoids. *Crop Protection* 20, 779–799.

Gold, C.S., Pinese, B., and Peña, J.E. (2002) Pests of banana. In: Peña, J.E., Sharp J.L., and Wysoki, M. (eds) *Tropical Fruit Pests and Pollinators. Biology, Economic Importance, Natural Enemies and Control*. CAB International, Wallingford, UK, pp. 13–56.

Gonzalez-Hernandez, H. (1995) The status of the biological control of pineapple mealybugs in Hawaii. PhD thesis, The University of Hawaii at Manoa, Honolulu, Hawaii, 128 pp.

Greathead, D.J. (1992) Natural enemies of tropical locusts and grasshoppers: their impact and potential as biological control agents. In: Lomer, C.J. and Prior, C. (eds) *Biological Control of Locusts and Grasshoppers*, CAB International, Wallingford, UK ,pp. 105–121.

Hazarika, L.K., Bhuyan, M., and Hazarika, B. (2008) Insect pests of tea and their management. *Annual Review of Entomology* 54, 267–284.

Herren, H.R. and Neuenschwander, P. (1991) Biological control of cassava pests in Africa. *Annual Review of Entomology* 36, 257–283.

Hill, D.S. (1975) *Agricultural Insect Pests of the Tropics and their Control*. Cambridge University Press, Cambridge, 516 pp.

Hill, D.S. (1987) *Agricultural Insect Pests of Temperate Regions and their Control*. Cambridge University Press, Cambridge, 659 pp.

Hokkanen, H. and Pimentel, D. (1984) New approach for selecting biological control agents. *The Canadian Entomologist* 116, 1109–1121.

IITA (2010) *Joining Hands Against Pod Borer*. IITA, R4D Review, Nigeria, 57 pp.

IPM CRSP (2009) *Annual Report FY 2009*. OIRED, Virginia Tech, Virginia, 365 pp.

IPM CRSP (2010) *Annual Report FY 2010*. OIRED, Virginia Tech, Virginia, 177 pp.

Jamieson, L.E., Froud, K., Edwards, R., and Stevens, P.S. (2008) Establishment of *Thripobius javae* (=*semiluteus*) in New Zealand. The New Zealand Plant Protection Society 61st Annual Conference, Paihia, Bay of Islands, New Zealand, 12–14 August 2008, pp. 17–23.

Kakimoto, K., Inoue, H., and Kashio, T. (2008) Effect of granular insecticides on *Thrips palmi*

Karny and its predatory natural enemies, *Orius strigicollis* (Poppius) and *Haplothrips brevitubus* (Karny). *Kyushu Plant Protection Research* 54, 85–92.

Kalshoven, L.G.E. (1981) *The Pests of Crops in Indonesia* (revised by P.A. Van der Laan). P.T. Ichtiar Baru – Van Hoeve, Jakarta, Indonesia, 701 pp.

Khoo, K.C., Ooi, P.A.C., and Ho, C.T. (1991) *Crop Pests and Their Management in Malaysia.* Tropical Press Sdn. Bhd., Kuala Lumpur, 242 pp.

King, A.B.S. and Saunders, J.L. (1984) *Las Plagas Invertebradas de Cultivos Anuales Alimenticios en América Central.* Overseas Development Administration, London, 182 pp.

Kogan, M., Gerling, D., and Maddox, J.V. (1999) Enhancement of biological control in annual agricultural environments. In: Bellows, T.S. and Fisher, T.W. (eds) *Handbook of Biological Control.* Academic Press, San Diego, pp. 789–818.

Kumar, N.S. and Ananthakrishnan, T.N. (1984) Predator-thrips interactions with reference to *Orius maxidentex* Ghauri and *Carayonocoris indicus* Muraleedharan (Anthocoridae: Hemiptera). *Proceedings of the Indian National Science Academy* B50, 139–145.

Lapis, E.B. (1970) The biology of the grey mealybug, *Ferrisia virgata* (Cockerell) (Pseudococcidae, Homoptera). *Philippine Entomologist* 21, 397–405.

Le Pelley, R.H. (1968) *Pests of Coffee.* Longmans, London, 590 pp.

Lewis, T. (1997) *Thrips as Crop Pests.* CAB International, Wallingford, UK, 740 pp.

Lin, Q.Y. (2005) Selection and application of *Beauveria bassiana* to control *Tessaratoma papillosa. Forest Research* 18, 143–146.

Liu Y.F., Zhang, G.R., Gu, D.X., and Zhang, W.Q. (2000) The natural control effect of egg parasitoids on litchi stinkbug *Tessaratoma papillosa* (Heteroptera: Pentatomidae) in Litchi orchard. *Entomologia Sinica* 7, 322–328.

Muniappan, R., Lawrence, J.H., and Marler, T.E. (2000) Greenhouse and red banded thrips and two of their natural enemies on *Elaeocarpus joga* in Guam. *Journal of Hawaiian Pacific Agriculture* 11, 61–63.

Muniappan, R., Shepard, B.M., Watson, G.W., Carner, G.R., Sartiami, D., Rauf, A., and Hammig, M.D. (2009) First report of the papaya mealybug, *Paracoccus marginatus* (Hemiptera: Pseudococcidae), in Indonesia and India (2008). *Journal of Agricultural and Urban Entomology* 25, 37–40.

Muniappan, R., Shepard, B.M., Watson, G.W., Carner, G.R., Rauf, A., Sartiami, D., Hidayat, P., Afun, J.V.K., Goergen, G., and Ziaur Rahman, A.K.M. (2010) New records of invasive insects (Hemiptera: Sternorrhyncha) in Southeast Asia and West Africa (2009). *Journal of Agricultural and Urban Entomology* 26, 167–173.

Nanta, P., Morakote, R., and Sirising, S. (1995) Biological control of agricultural insect pests in Thailand. In: Ferrar, P., Loke, W.H., Sivapragasam, A., Ismail, A.A., Lum, K.Y., Lim, G.S. and Vos, J. (eds) *Biological Control as a Cornerstone of Integrated Pest Management for Sustainable Agriculture in Southeast Asia,* MARDI and ACIAR, pp. 85–95.

Navasero, M.V. (1983) Biology and chemical control of the eggplant fruit and shoot borer, *Leucinodes orbonalis* Guenée (Pyraustidae: Lepidoptera). BS Thesis, University of the Philippines at Los Baños, Philippines.

Nayar, K.K, Ananthakrishnan, T.N., and David, B.V. (1976) *General and Applied Entomology.* Tata McGraw-Hill Publishing Company Limited, New Delhi, 589 pp.

Neuenschwander, P., Borgemeister, C., and Langewald, J. (2003) *Biological Control in IPM Systems in Africa.* CAB International, Wallingford, UK, 414 pp.

Obinna, A., Neuenschwander, P., and Korie, S. (2010) Niche separation between *Encarsia dispersa* and *Encarsia guadeloupe,* two biological control agents of the spiraling whitefly *Aleurodicus dispersus,* in Benin, West Africa. *BioControl* DOI 10.1007/s10526-010-9331-9.

O'Connor, B.A. (1969) *Exotic Plant Pests and Diseases. A Handbook of Plant Pests and Diseases to be Excluded from or Prevented from Spreading within the Area of the South Pacific Commission.* South Pacific Commission, Noumea, New Caledonia.

Ooi, P.A.C. (1988) *Insects in Malaysian Agriculture*. Tropical Press Sdn. Bhd., Kuala Lumpur, 106 pp.

Ooi, P.A.C., Winotai, A., and Peña, J.E. (2002) Pests of minor tropical fruits. In: Peña, J.E., Sharp, J.L., and Wysoki, M. (eds.) *Tropical Fruit Pests and Pollinators: Biology, Economic Importance, Natural Enemies and Control*. CAB International, Wallingford, UK, pp. 315–330.

Paine, R.W. (1964) *The Banana Scab Moth,* Nacoleia octasema *(Meyrick): Its Distribution, Ecology and Control.* South Pacific Commission, Noumea, New Caledonia, Technical Paper No. 145, 70 pp.

Pappu, H.R., Jones, R.A.C., and Jain, R.K. (2009) Global status of tospovirus epidemics in diverse cropping systems: Successes achieved and challenges ahead. Virus Research Doi: 10.1016/j.virusres.2009.01.009.

Parker, B.L., Talekar, N.S., and Skinner, M. (1995) *Field Guide: Insect Pests of Selected Vegetables in Tropical and Subtropical Asia*. AVRDC Publication, pp. 94–427.

Paull, R.E. and Duarte, O. (2011) *Tropical Fruits Volume 1*, 2nd Edition. CAB International, Wallingford, UK, 400 pp.

Peña, J.E., Sharp, J.L., and Wysoki, M. (2002) *Tropical Fruit Pests and Pollinators: Biology, Economic Importance, Natural Enemies and Control*. CAB International, Wallingford, UK, 430 pp.

Petty, G.J., Stirling, G.R., and Bartholomew, D.P. (2002) Pests of pineapple. In: Peña, J.E., Sharp, J.L., and Wysoki, M. (eds) *Tropical Fruit Pests and Pollinators: Biology, Economic Importance, Natural Enemies and Control*. CAB International, Wallingford, UK, pp. 157–195.

Prior, C. (1992) Discovery and characterization of fungal pathogens for locust and grasshopper control. In: Lomer, C.J., and Prior, C. (eds) *Biological Control of Locusts and Grasshoppers,* CAB International, Wallingford, UK, pp. 159–180.

Reddy, G.V.P., Muniappan, R., Cruz, Z.T., Naz, F., Bamba, J.P., and Tenorio, J. (2009) Present status of *Maconellicoccus hirsutus* (Hemiptera: Pseudococcidae) in the Mariana Islands and its control by two fortuitously introduced natural enemies. *Journal of Economic Entomology* 102, 1431–1439.

Richards, A.M. (1983) The *Epilachna vigintioctopunctata* complex (Coleoptera: Coccinellidae). *International Journal of Entomology* 25, 11–41.

Sertkaya, E., Telli, S., and Yiğit, A. (2006) Pest status and parasitoids of Gall midge, *Asphondylia capsici* Barnes (Diptera: Cecidomyiidae) in Antakya province. *Türkiye Entomoloji Dergisi* 30, 223–234.

Sharma, H.C. (1998) Bionomics, host plant resistance, and management of the legume pod borer, *Maruca vitrata* – a review. *Crop Protection* 17, 373–386.

Shepard, B.M. and Barrion, A.T. (1998) Parasitoids of insects associated with soybean and vegetable crops in Indonesia. *Journal of Agricultural Entomology* 15, 239–272.

Shepard, B.M., Carner, G.R., Barrion, A.T., Ooi, P.A.C., and van den Berg, H. (1999) *Insects and their Natural Enemies Associated with Vegetables and Soybean in Southeast Asia*. Quality Printing Company, Orangeburg, SC, USA, 108 pp.

Smith, D. and Peña, J.E. (2002) Tropical citrus pests. In: Peña, J.E., Sharp, J.L., and Wysoki, M. (eds) *Tropical Fruit Pests and Pollinators: Biology, Economic Importance, Natural Enemies, and Control.* CAB International, Wallingford, UK, pp. 57–101.

Smith, R.I. (1977) *Cosmopolites sordidus* Germ. In: Kranz, J., Schmutterer, H., and Koch, W. (eds) *Diseases, Pests and Weeds in Tropical Crops*. John Wiley & Sons, New York, pp 406–410.

SP-IPM (2010) *Integrated Pest Management and Crop Health – Bringing Together Sustainable Agroecosystems and People's Health.* White Paper, SP-IPM Secretariat, International Institute of Tropical Agriculture (IITA), Ibadan, Nigeria, 17 pp.

Sran, C.S. and Sandhu, G.S. (1979) New records of parasites and predators of *Nephopteryx eugraphella* Ragonot in India. *Current Science* 48, 169.

Srinivasan, R. (2008) Integrated pest management for eggplant fruit and shoot borer (*Leucinodes orbonalis*) in South and Southeast Asia: past, present and future. *Journal of Biopesticides* 1, 105–112.

Srinivasan, R. (2009) *Insect and Mite Pests on Eggplant: A Field Guide for Identification and Management*. AVRDC – The World Vegetable Center, Shanhua, Taiwan, AVRDC Publication No. 09-729, 64 pp.

Sundararaju, D. and Sundara Babu, P.C. (1999) *Helopeltis* spp. (Heteroptera: Miridae) and their management in plantation and horticultural crops of India. *Journal of Plantation Crops* 27, 155–174.

Tamò, M., Ekesi, S., Maniania, N.K., and Cherry, A. (2003) Biological control, a non-obvious component of IPM for cowpea, In: Neuenchwander, P., Borgemeister, C. and Langewald, J. (eds) *Biological Control in IPM Systems in Africa*. CAB International, Wallingford, UK. pp. 311–328.

Vega, F.E. (2008) The rise of coffee. *American Scientist* 96, 138–145.

Vega, F.E., Infante, F., Castillo, A., and Jaramillo, J. (2009) The coffee berry borer, *Hypothenemus hampei* (Ferrari) (Coleoptera: Scolytidae): a short review, with recent findings and future research directions. *Terrestrial Arthropod Reviews* 2, 129–147.

Waite, G.K. (2002) Pests and pollinators of mango. In: Peña, J.E., Sharp, J.L., and Wysoki, M. (eds) *Tropical Fruit Pests and Pollinators: Biology, Economic Importance, Natural Enemies and Control*. CAB International, Wallingford, UK, pp. 103–129.

Waite, G.K. and Gerson, U. (1994) The predator guild associated with *Aceria litchii* (Acari: Eriophyidae) in Australia and China. *Entomophaga* 39, 275–80.

Waite, G.K. and Hwang, J.S. (2002) Pests of litchi and longan. In: Peña, J.E., Sharp, J.L., and Wysoki, M. (eds) *Tropical Fruit Pests and Pollinators: Biology, Economic Importance, Natural Enemies and Control*. CAB International, Wallingford, UK, pp. 331–359.

Wang, Y., Watson, G.W., and Zhang, R. (2010) The potential distribution of an invasive mealybug *Phenacoccus solenopsis* and its threat to cotton in Asia. *Agricultural and Forest Entomology* DOI: 10.1111/j.1461-9563.2010.00490.x.

Waterhouse, D.F. (1993a) *Biological Control: Pacific Prospects – Supplement 2*. ACIAR Monograph No. 20, Australian Centre for International Agricultural Research (ACIAR), 138 pp.

Waterhouse, D.F. (1993b) *The Major Arthropod Pests and Weeds of Agriculture in Southeast Asia*, Australian Centre for International Agricultural Research, Canberra, Australia, 141 pp.

Waterhouse, D.F. (1998) *Biological Control of Insect Pests: Southeast Asia Prospects*. Australian Centre for International Agricultural Research, Canberra, Australia, 548 pp.

Waterhouse, D.F. and Norris, K.R. (1987) *Biological Control: Pacific Prospects*. Inkata Press, Melbourne, Australia, 454 pp.

Waterhouse, D.F. and Norris, K.R. (1989) *Biological Control: Pacific Prospects – Supplement 1*, Australian Centre for International Agricultural Research, Canberra, Australia, 123 pp.

Waterhouse, G.F. and Sands, D.P.A. (2001) *Classical Biological Control of Arthropods in Australia*. ACIAR Monograph No. 77, 560 pp.

Watson, G.E., Ooi, P.A.C., and Girling, D.J. (1995) *Insects on Plants in the Maldives and their Management*. International Institute of Biological Control, Ascot, U.K., 124 pp.

Winotai, A., Goergen, G., Tamo, M., and Neuenschwander, P. (2010) Cassava mealybug has reached Asia. *Biocontrol News and Information* 31, 10N–11N.

Wu, W.N., Lan, W.M., and Liu, Y.H. (1991) Phytoseiid mites on litchis in China and their application. *Natural Enemies of Insects* 13, 82–91.

Wysoki, M., van den Berg, M.A., Ish-Am, G., Gazit, S., Peña, J.E., and Waite, G.K. (2002) Pests and pollinators of avocado. In: Peña, J.E., Sharp, J.L., and Wysoki, M. (eds) *Tropical Fruit Pests and Pollinators*. CAB International, Wallingford, UK, pp. 223–293.

Yaacob, O. and Subhadrabandhu, S. (1995) *The Production of Economic Fruits in South-East Asia.* Oxford University Press, Kuala Lumpur, Malaysia, 419 pp.

Yasodha, P. and Natarajan, N. (2006) Diversity of natural enemies of *Leucinodes orbonalis* Guenée (Lepidoptera: Pyraustidae). *Entomon* 31, 323–326.

Yunus, A. and Balasubramaniam, A. (1981) *Major Crop Pests in Peninsular Malaysia* (Second edition). Ministry of Agriculture, Peninsular Malaysia, Kuala Lumpur, 190 pp.

Yunus, A. and Ho, T.H. (1980) *List of Economic Pests, Host Plants, Parasites and Predators in West Malaysia (1920–1978).* Ministry of Agriculture Bulletin No. 153, Kuala Lumpur, 538 pp.

List of Parasitoids and Predators

Ablerus elegantulus (Silvestri) (Hym.: Aphelinidae)
Aceratoneuromyia indica (Silvestri) (Hym.: Eulophidae)
Acerophagus papayae Noyes and Schauff (Hym.: Encyrtidae)
Actia aegyptia Vill. (Dip.: Tachinidae)
Actia hyalinata Malloch (Dip.: Tachinidae)
Adalia bipunctata (L.) (Col.: Coccinellidae)
Adonia variegata (Goeze) (Col.: Coccinellidae)
Aenasius advena Compere (Hym.: Encyrtidae)
Aenasius bambawale Hayat (Hym.: Encyrtidae)
Agathis sp. (Hym.: Braconidae)
Ageniaspis citricola Logvinovskaya (Hym.: Encyrtidae)
Agiommatus sp. nr. *sumatraensis* Crawford (Hym.: Pteromalidae)
Agistemus exsertus Gonzalez (Acari: Stigmaeidae)
Alaphora sp. (Dip.: Tachinidae)
Aleochara sp. (Col. Staphylinidae)
Aleurodothrips fasciapennis (Franklin) (Thy.: Phlaeothripidae)
Allograpta nasuta (Macquart) (Dip.: Syrphidae)
Allograpta obliqua Say (Dip.: Syrphidae)
Allotropa sp. nr. *mecrida* (Walker) (Hym.: Platygasteridae)
Amblyseius benjamini Schicha (Acari: Phytoseidae)
Amblyseius cucumeris (Oudemans) (Acari: Phytoseiidae)
Amblyseins largoensis (Muma) (Acari: Phytoseiidae)
Amblyseius mckenziei Sch.& Pr. (Acari: Phytoseiidae)
Amblyseius spp. (Acari: Phytoseiidae)
Amitus hesperidum Silvestri (Hym.: Aphelinidae)
Amobia sp. (Diptera: Sarcophagidae)
Amyosoma leuzerae Rohwer (Hym.: Braconidae)
Anagyrus ananatis Gahan (Hym.: Encyrtidae)
Anagyrus flaveolus Waterhouse (Hym.: Encyrtidae)
Anagyrus greeni Howard (Hym.: Encyrtidae)
Anagyrus kamali Moursi (Hym.: Encyrtidae)

Anagyrus loecki Noyes (Hym.: Encyrtidae)
Anagyrus mangicola Noyes (Hym.: Encyrtidae)
Anagyrus sp. (Hym.: Encyrtidae)
Anastatus japonicus Ashmead (Hym.: Eupelmidae)
Anastatus sp. (Hym.: Eupelmidae)
Anicetus beneficus Ishii & Yasumatsu (Hym.: Encyrtidae)
Anoplolepis longipes (Jerdon) (Hym.: Formicidae)
Anthocoris spp. (Hem.: Anthocoridae)
Antilochus coquebertii (Fabricius) (Hem.: Pyrrhocoridae)
Antrocephalus nr. *subelongatus* Kohl (Hym.: Chalcididae)
Apanteles antipoda Ashmead (Hym.: Braconidae)
Apanteles crocidolomiae Rislec (Hym.: Braconidae)
Apanteles erionotae Wilkinson (Hym.: Braconidae)
Apanteles etiellae isolatus Muesebeck (Hym.: Braconidae)
Apanteles glomeratus (Linnaeus) (Hym.: Braconidae)
Apanteles machaeralis Wilkinson (Hym.: Braconidae)
Apanteles mendosae Wilkinson (Hym.: Braconidae)
Apanteles obliquae Wilkinson (Hym.: Braconidae)
Apanteles papilionis Vier. (Hym.: Braconidae)
Apanteles parasae Rohwer (Hym.: Braconidae)
Apanteles ruidus Wilkinson (Hym.: Braconidae)
Apanteles sp. (Hym.: Braconidae)
Apanteles spp. (Hym.: Braconidae)
Apanteles subandinus Blanchard (Hym.: Braconidae)
Apanteles sundanus Wilkinson (Hym.: Braconidae)
Apanteles taragamae Viereck (Hym.: Braconidae)
Aphaereta pallipes (Say) (Hym.: Braconidae)
Aphelinus abdominalis (Dalmer) (Hym.: Aphelinidae)
Aphelinus gossypii Timberlake (Hym.: Aphelinidae)
Aphelinus mali (Haldeman) (Hym.: Aphelinidae)
Aphidencyrtus sp. (Hym.: Encyrtidae)
Aphidius absinthii Marshall (Hym.: Aphidiidae)
Aphidius colemani Viereck (Hym.: Aphidiidae)
Aphidius gifuensis Ashmead (Hym.: Aphidiidae)
Aphidius matricariae (Haliday) (Hym.: Aphidiidae)
Aphidius similis Starý and Carver (Hym.: Aphidiidae)
Aphidius spp. (Hym.: Aphidiidae)
Aphidius uzbekistanicus Luzhetzki (Hym.: Aphidiidae)
Aphytis chrysomphali (Mercet) (Hym.: Aphelinidae)
Aphytis lingnanensis Compere (Hym.: Aphelinidae)
Aphytis sp. (Hym.: Aphelinidae)
Apoanagyrus californicus Compere (Hym.: Encyrtidae)
Apophua sp. (Hym.: Ichneumonidae)
Aprostocetus ceroplastae (Girault) (Hym.: Eulophidae)
Aprostocetus sp. (Hym.: Eulophidae)
Argiope spp. (Arachnida: Aranidae)
Argyrophylax (*Sturmia*) *albincisa* (Wiedemann) (Dip.: Tachinidae)
Asecodes delucchi (Boucek) (Hym.: Eulophidae)
Aspergillus origer (Fungus)
Aspidiotiphagus sp. (Hym.: Encyrtidae)
Atherigona orientalis Schiner (Dip.: Muscidae)

Baeognatha (= *Agathis*) sp. (Hym.: Braconidae)
Bardylis parvipennis (Gahan) (Hym.: Encyrtidae)
Bassus bruesi Shenefelt (Hym.: Braconidae)
Bassus cylasovorus Rohwer (Hym.: Braconidae)
Bassus javanicus (Bhat and Gupta) (Hym.: Braconidae)
Bassus sp. (Hym.: Braconidae)
Bdella distincta Baker and Balock (Acari.: Bdellidae)
Beauveria bassiana (Balsamo) (Fungus)
Belonuchus ferrugatus (Erichson) (Col.: Staphylinidae)
Bessa remota (Aldrich) (Dip.: Tachinidae)
Betasyrphus adligatus Wiedemann (Diptera: Syrphidae)
Biosteres persulcatus Silvestri (Hym.: Braconidae)
Blastophaga sp. (Hym.: Agaonidae)
Bleparipa (= *Sturmia*) sp. (Dip.: Tachinidae)
Blepharella lateralis Macquart (Dip.: Tachinidae)
Blepyrus insularis (Cameron) (Hym.: Encyrtidae)
Blondelia (*Eucelatoria*) *armigera* (Coq.) (Dip.: Tachinidae)
Brachymeria albotibialis (Ashmead) (Hym.: Chalcididae)
Brachymeria bengalensis pulchellae Joseph, Narendran and Joy (Hym.: Chalcididae)
Brachymeria euploeae (Westwood) (Hym.: Chalcididae)
Brachymeria lasus (Walker) (= *Brachymeria obscurata* Walker) (Hym.: Chalcididae)
Brachymeria marginata Cameron (Hym.: Chalcididae)
Brachymeria nephantidis Gahan (Hym.: Chalcididae)
Brachymeria ovata (Say) (Hym.: Chalcididae)
Brachymeria sp. (Hym.: Chalcididae)
Brachymeria sp. nr. *marginicollis* (Cambridge) (Hym.: Chalcididae)
Bracon brevicornis Wesmael (Hym.: Braconidae)
Bracon cajani Muesebeck (Hym.: Braconidae)
Bracon cylasovorus (Rohwer) (Hym.: Braconidae)
Bracon greeni Ashmead (Hym.: Braconidae)
Bracon hebetor Say (Hym.: Braconidae)
Bracon kirkpatricki Wilkinson (Hym.: Braconidae)
Bracon lefroyi (Dudgeon and Gough) (Hym.: Braconidae)
Bracon melleus Ayyer (Hym.: Braconidae)
Bracon pectoralis Wesmael (Hym.: Braconidae)
Bracon piger Wesmael (Hym.: Braconidae)
Bracon sp. (Hym.: Braconidae)
Bracon thurberiphagae (Muesebeck) (Hym.: Braconidae)
Braunsia kriegeri Enderlein (Hym.: Braconidae)
Brumus sp. (Col.: Coccinellidae)
Brumus suturalis (F.) (Col.: Coccinellidae)
Cadurcia sp. (Dip.: Tachinidae)
Caenoplana coerulea Moseley (Platyhelminthes: Geoplanidae)
Calleida splendidula (F.) (Col.: Carabidae)
Calliephialtes sp. (Hym.: Ichneumonidae)
Camponotus compressus (F.) (Hym.: Formicidae)
Carayonocoris indicus Muraleedharan (Hem.: Anthocoridae)
Carcelia buitenzorgiensis Bar. (Tachinidae)
Cathartus sp. (Col.: Silvanidae)
Cedria paradoxa Wilkinson (Hym.: Braconidae)
Centrochalcis sp. (Hym.: Chalcididae)

Centrodora sp. (Hym.: Aphelinidae)
Centrodora sp. nr. *scolypopae* Valentine, (Hym.: Aphelinidae)
Cephalonomia stephanoderis Betrem (Hym.: Bethylidae)
Cephalosporium crassum (Fungus)
Ceranisus americenis (Girault) (Hym.: Eulophidae)
Ceranisus menes Walker (Hym.: Eulophidae)
Ceranisus russelli De Santis (Hym.: Eulophidae)
Ceranisus vinctus (Gahan) (Hym.: Eulophidae)
Ceraphron sp. (Hym.: Ceraphronidae)
Ceratomegilla maculata De Geer (Col.: Coccinellidae)
Chaetogaedia monticola (Big.) (Dip.: Tachnidae)
Chalcis sp. (Hym.: Chalcididae)
Chalcis tachardiae Cameron (Hym.: Chalcididae)
Charops bicolor (Szepligeti) (Hym.: Ichneumonidae)
Charops brachypterum Gupta and Maheswary (Hym.: Ichneumonidae)
Charops sp. (Hym.: Ichneumonidae)
Cheilomenes septempunctata L. (Col.: Coccinellidae)
Cheilomenes vicina (Mulsant) (Col.: Coccinellidae)
Chelisoches morio (Fabricius) (Der.: Chelisochidae)
Chelonus busckiella Viereck (Hym.: Braconidae)
Chelonus inanitus (Linnaeus) (Hym.: Braconidae)
Chelonus insularis (Cresson) (Hym.: Braconidae)
Chelonus rufus Lyle (Hym.: Braconidae)
Chelonus sp. (Hym.: Braconidae)
Chelonus striatigenas Cameron (Hym.: Braconidae)
Chilocorus cacti (Linnaeus) (Col.: Coccinellidae)
Chilocorus malasiae Crotch (Col.: Coccinellidae)
Chilocorus nigritus (Fabricius) (Col.: Coccinellidae)
Chilocorus politus Mulsant (Col.: Coccinellidae)
Chlorocytus sp. (Hym.: Pteromalidae)
Chrysochalcissa oviceps Boucek (Hym.: Torymidae)
Chrysocharis pentheus (Walker) (Hym.: Eulophidae)
Chrysopa spp. (Neu.: Chrysopidae)
Chrysoperla carnea (Stephens) (Neu.: Chrysopidae)
Chrysopilus ferruginosus Wiedermann (Dip.: Rhagionidae)
Citrostichus phyllocnistoides (Narayanan) (Hym.: Eulophidae)
Closterocerus trifasciatus Westwood (Hym.: Eulophidae)
Coccidencyrtus ochraceipes Gahan (Hym.: Encyrtidae)
Coccidoctonus dubius (Girault) (Hym.: Encyrtidae)
Coccidodiplosis smithi (Felt) (Dip.: Cecidomyiidae)
Coccinella septempunctata L. (Col.: Coccinellidae)
Coccinella septempunctata brucki Mulsant (Col.: Coccinellidae)
Coccinella transversalis Fabricius (= *Coccinella repanda* Thunberg) (Col.: Coccinellidae)
Coccophagus ceroplastae (Howard) (Hym.: Aphelinidae)
Colpotrichia sp. (Hym.: Ichneumonidae)
Comperiella bifasciata Howard (Hym.: Encyrtidae)
Comperiella unifasciata Ishii (Hym.: Encyrtidae)
Copidosoma koehleri Blanchard (Hym.: Encyrtidae)
Copidosoma sp. (Hym.: Encyrtidae)
Copidosomopsis sp. (Hym.: Encyrtidae)
Cordyceps coccinea (Fungus)

Cotesia (*Apanteles*) *erionotae* (Wilkinson) (Hym.: Braconidae)
Cotesia marginiventris (Cresson) (Hym.: Braconidae)
Cotesia plutellae Kurdyumov (Hym.: Braconidae)
Cotesia sp. (Hym.: Braconidae)
Cremastus hapaliae Cushman (Hym.: Ichneumonidae)
Cremastus sp. (Hym.: Ichneumonidae)
Crematogaster sp. (Hym.: Formicidae)
Cremnops sp. (Hym.: Braconidae)
Creobroter gemmatus Stol (Ort.: Mantidae)
Cristicaudus nepalensis (Takada) (Hym.: Aphidiidae)
Cryptognatha gemellata Mulsant (Col.: Coccinellidae)
Cryptognatha nodiceps Marshall (Col.: Coccinellidae)
Cryptogonus orbiculus nigripennis Weise (Col.: Coccinellidae)
Cryptolaemus montrouzieri Mulsant (Col.: Coccinellidae)
Curinus coeruleus (Mulsant) (Col.: Coccinellidae)
Cycloneda sanguineda (L.) (Col.: Coccinellidae)
Dacnusa sibirica Telenga (Hym.: Braconidae)
Dactylosternum hydrophiloides Macleay (Col.: Hydrophilidae)
Delphastus spp. (Col.: Coccinellidae)
Diachasmimorpha dausii (Cameron) (Hym.: Braconidae)
Diachasmimorpha hageni (Fullaway) (Hym.: Braconidae)
Diachasmimorpha longicaudata (Ashmead) (Hym.: Braconidae)
Diadegma apostata (Gravenhorst) (Hym.: Ichneumonidae)
Diadegma eucerophaga Horstmann (Hym.: Ichneumonidae)
Diadegma fenestralis (Holmgr.) (Hym.: Ichneumonidae)
Diadegma pulchripes Kokujev (Hym.: Ichneumonidae)
Diadegma semiclausum (Hellen) (Hym.: Ichneumonidae)
Diadegma sp. (Hym.: Ichneumonidae)
Diadromus collaris (Gravenhorst) (Hym.: Ichneumonidae)
Diaeretiella rapae (M'Intosh) (Hym.: Braconidae)
Diaglytidea roepkei (Viereck) (Hym.: Ichneumonidae)
Diaphorencyrtus aligarhensis (Shafee, Alam, and Agarwal) (Hym.: Encyrtidae)
Diglyphus isaea (Walker) (Hym.: Eulophidae)
Dinarmus sp. (Hymenoptera: Pteromalidae)
Diplazon orientalis (Cameron) (Hym.: Ichneumonidae)
Dirhinus anthracina Walker (Hym.: Chalcididae)
Diversinervis elegans Silvestri (Hym.: Encyrtidae)
Dolichoderus thoracicus (Smith) (Hym.: Formicidae)
Dolichogenidea sp. (Hym.: Braconidae)
Dolichogenidea stantoni (Ashmead) (Hym.: Braconidae)
Doryctes bistriatus Kieffer (Hym.: Braconidae)
Doryctes brevipetiolus Kieffer (Hym.: Braconidae)
Doryctes picticeps Kieffer (Hym.: Braconidae)
Doryctes strioliger Kieffer (Hym.: Braconidae)
Doryctes tristriatus Kieffer (Hym.: Braconidae)
Echthromorpha agrestoria (Swederus) (Hym.: Ichneumonidae)
Echthromorpha striata Kriegler (Hym.: Ichneumonidae)
Eiphosoma dentator (Fabricius) (Hym.: Ichneumonidae)
Elasmus indicus Rowher (Hym.: Elasmidae)
Elasmus johnstoni Ferrière (Hym.: Elasmidae)
Elasmus philippinensis Ashmead (Hym.: Elasmidae)

Elasmus sp. (Hym.: Elasmidae)
Encarsia citrina (Crawford) (Hym.: Aphelinidae)
Encarsia clypealis (Silvestri) (Hym.: Aphelinidae)
Encarsia dispresa Polaszek (Hym.: Aphelinidae)
Encarsia guadeloupe Viggiani (Hym.: Aphelinidae)
Encarsia haitiensis Dozier (Hym.: Aphelinidae)
Encarsia opulenta (Silvestri) (Hym.: Aphelinidae)
Encarsia smithi (Silvestri) (Hym.: Aphelinidae)
Encarsia sp. (Hym.: Aphelinidae)
Encarsia spp. (Hym.: Aphelinidae)
Encyrtus (*Ooencyrtus*) sp. (Hym.: Encyrtidae)
Enicospilus atricornis (Morley) (Hym.: Ichneumonidae)
Enicospilus sp. (Hym.: Ichneumonidae)
Enicospilus xanthocephalus Cameron (Hym.: Ichneumonidae)
Entomophthora fresenii Nowakowski (Fungus)
Eocanthecona furcellata (Wolff) (Hem.: Pentatomidae)
Ephedrus persicae Froggatt (Hym.: Aphidiidae)
Ephedrus plagiator (Nees) (Hym.: Aphidiidae)
Ephedrus sp. nr *cerasicola* Stary (Hym.: Aphidiidae)
Epipyrops fuliginosa Tams (Lep.: Epipyropidae)
Episyrphus balteatus (De Geer) (Dip.: Syrphidae)
Epixorides caerulescens Morley (Hym.: Ichneumonidae)
Eretmocerus spp. (Hym.: Aphelinidae)
Eretmocerus serius Silvestri (Hym.: Aphelinidae)
Eriborus argenteopilosus (Cameron) (Hym.: Ichneumonidae)
Eriborus sinicus (Holmgren) (Hym.: Ichneumonidae)
Erycia nymphalidaephaga Bar. (Hym.: Ichneumonidae)
Euborellia annulata (Lucas) (Dem.: Carcinophoridae)
Euderus malayensis Ferrière (Hym.: Eulophidae)
Euderus sp. (Hym.: Eulophidae)
Eumenes petiolata (F.) (Hym.: Vespidae)
Eupelmus urozonus Dalman (Hym.: Eupelmidae)
Eupeodes confrater (Wiedemann) (Dip.: Syrphidae)
Euplectrus cf. *chapadae* (Asmead) (Hym.: Eulophidae)
Euplectrus leucostomus Rohwer (Hym.: Eulophidae)
Euplectrus manilae Ash. (Hym.: Eulophidae)
Euplectrus maternus Bhatnagar (Hym.: Eulophidae)
Euplectrus platyhypenae Ashmead (Hym.: Eulophidae)
Euplectrus sp. (Hym.: Eulophidae)
Euplectrus spp. (Hym.: Eulophidae)
Euplectrus taiwanus Sonan (Hym.: Eulophidae)
Euryrhopalus propinquus Kerrich (Hym.: Encyrtidae)
Eurytoma dentata Mayr (Hym.: Eurytomidae)
Eurytoma poloni Girault (Hym.: Eurytomidae)
Eurytoma sp. (Hym.: Eurytomidae)
Eurytoma xylotrechi (Hym.: Eurytomidae)
Exochus sp. (Hym.: Braconidae)
Exorista sorbillans (Wied.) (Dip.: Tachinidae)
Exorista sp. (Diptera: Tachinidae)
Fopius arisanus (Sonan) (Hym.: Braconidae)
Fopius sp. (Hym.: Braconidae)

Fopius vandenboschi (Fullaway) (Hym.: Braconidae)
Franklinothrips megalops Trybom (Thy.: Aeolothripidae)
Franklinothrips tenuicornis Hood (Thy.: Aeolothripidae)
Franklinothrips vespiformis (Crawford) (Thy.: Aeolothripidae)
Fusarium sp. (Fungus)
Geocoris spp. (Hem.: Lygaeidae)
Glyptapanteles phytometrae (Wilkinson) (Hym.: Braconidae)
Glyptomorpha sp. (Hym.: Braconidae)
Goetheana (*Dasyscapus*) *parvipennis* (Gahan) (Hym.: Eulophidae)
Goetheana shakespearei Girault (Hym.: Eulophidae)
Gonatocerus sp. (Hym.: Mymaridae)
Goniozus (= *Perisierola*) sp. (Hym.: Bethylidae)
Goniozus sensorius Gordh (Hym.: Bethylidae)
Goryphus fasciatipennis Szepl. (Hym.: Ichneumonidae)
Goryphus javanicus (Roman) (= *B. mesoxanthus* Br.) (Hym.: Ichneumonidae)
Goryphus nursei (Cameron) (Hym.: Ichneumonidae)
Goryphus sp. (Hym.: Ichneumonidae)
Granulosis virus
Gryon sp. (Hym.: Scelionidae)
Gyranusoidea indica Shafee, Alam, & Agarwal (Hym.: Encyrtidae)
Gyranusoidea tebygi Noyes (Hym.: Encyrtidae)
Hadrojoppa cognatoria Smith (Hym.: Ichneumonidae)
Halticoptera ciculus (Walker) (Hym.: Pteromalidae)
Hambletonia pseudococcina Compere (Hym.: Encyrtidae)
Haplothrips brevitubus (Karny) (Thy.: Phlaeothripidae)
Harmonia dimidiata (F.) (Col.: Coccinellidae)
Hemiptarsenus sp. (Hym.: Eulophidae)
Hemiptarsenus varicornis (Girault) (Hym.: Eulophidae)
Heterospilus coffeicola Schmiedeknecht (Hym.: Braconidae)
Heterotylenchus aberrans Bovien (Nematoda: Parasitylenchidae)
Hippodomia convergens Guérin (Col.: Coccinellidae)
Hirsutella sp. (Fungus)
Hirsutella thompsonii (Fungus)
Holanusomyia pulchripennis Girault (Hym.: Encyrtidae)
Hololepta quadridentata (Fabricius) (Col.: Histeridae)
Hyposoter exguae (Vier.) (Hym.: Ichneumonidae)
Iphiaulax sp. (Hym.: Braconidae)
Iridomyrmex anceps (Roger) (Hym.: Formicidae)
Ischiodon scutellaris (Fabricius) (Dip.: Syrphidae)
Isaria sp. (Fungus)
Itamoplex sp. (Hym.: Ichneumonidae)
Kratoysma citri Bouček (Hym.: Eulophidae)
Leptomastix abyssinica Compere (Hym.: Encyrtidae)
Leptomastix dactylopii Howard (Hym.: Encyrtidae)
Lespesia archippivora (Riley) (Dip.: Tachinidae)
Leucopis sp. (Dip.: Chamaemyiidae)
Leurocerus hongkongensis Subba Rao (Hym.: Encyrtidae)
Lipolexis gracilis Förster (Hym.: Aphidiidae)
Lipolexis scutellaris Mackauer (Hym.: Aphidiidae)
Lissopimpla pacifica Marley (Hym.: Ichneumonidae)
Lixophaga sp. (Dip.: Tachinidae)

Lysaphidus schimitscheki (Fahriner) (Hym.: Aphidiidae)
Lysaphidus sp. (Hym.: Aphidiidae)
Lysiphlebia japonica (Ashmead) (Hym.: Aphidiidae)
Lysiphlebia mirzai Shuja Uddin (Hym.: Aphidiidae)
Lysiphlebus fabarum (Marshall) (Hym.: Aphidiidae)
Lysiphlebus sp. (Hym.: Aphidiidae)
Lysiphlebus testaceipes (Cresson) (Hym.: Aphidiidae)
Macrocentrus ancylivora Rohwer (Hym.: Braconidae)
Macrocentrus homonae Nixon (Hym.: Braconidae)
Macrocentrus nr. *trimaculatus* (Cameron) (Hym.: Braconidae)
Macromalon orientale Kerr. (Hym.: Ichneumonidae)
Marpissa sp. (Arachnida: Salticidae)
Mataeomera dubia Butler (Lep.: Noctuidae)
Megaphragma mymaripenne Timberlake (Hym.: Trichogrammatidae)
Megaselia chlumetiae Disney (Diptera: Phoridae)
Menochilus sexmaculatus (F.) (Col.: Coccinellidae)
Mesocomys orientalis Ferrière (Hym.: Eupelmidae)
Metapelma sp. (Hym.: Enpelmidae)
Metaphycus sp. (Hym.: Encyrtidae)
Metarhizium anisopliae (Metschnikoff) Sorokin (Fungus)
Metarhizium spp. (Fungus)
Metaseiulus occidentalis (Nesbitt) (Acari.: Phytoseiidae)
Meteorus autographae (Muesebeck) (Hym.: Braconidae)
Meteorus octasemae Fischer (Hym.: Braconidae)
Meteorus trichogrammae Wilkinson (Hym.: Braconidae)
Mexecheles hawaiiensis (Baker) (Acari: Cheyletidae)
Micraspis discolor (Fabricius) (Col.: Coccinellidae)
Microbracon recinicola (Hym.: Braconidae)
Microbracon thurberiphagae Muesebeck (Hym.: Braconidae)
Microdon bicolor Sack (Diptera: Syrphidae)
Microdus sp. (Hym.: Braconidae)
Micromeriella (*Campsomeris*) *marginella modesta* (Smith) (Hym.: Scoliidae)
Micromus timidus Hagen (Neu.: Hemerobiidae)
Microplitis maculipennis Szepligeti (Hym.: Braconidae)
Microplitis similes Lyle (Hym.: Braconidae)
Microplitis sp. (Hym.: Braconidae)
Microterys australicus Prinsloo (Hym.: Encyrtidae)
Microterys neitneri (Motschulsky) (Hym.: Encyrtidae)
Microtoridea lissonota Viereck (Hym.: Ichneumonidae)
Mirufens sp. nr. *mangiferae* Viggiani and Hayat (Hym.: Trichogrammatidae)
Monacrosporium ambrosium (Symbiotic fungus)
Moranila californica (Howard) (Hym.: Pteromalidae)
Moranila compere (Ashmead) (Hym.: Pteromalidae)
Mysepyris grandiceps Kieffer (Hym.: Bethylidae)
Nabis spp. (Hem.: Nabidae)
Nemorilla floralis Fallen (Dip.: Tachinidae)
Neochrysocharis formosa (Westwood) (Hym.: Eulophidae)
Neochrysocharis sp. (Hym.: Eulophidae)
Neopimploides syleptae Viereck (Hym.: Ichneumonidae)
Neoseiulus californicus (McGregor) (Acari: Phytoseiidae)
Neozygites parvispora (Fungus)

Nephaspis oculatus Blatchley (= *N. amnicola*) (Col.: Coccinellidae)
Nesolynx thymus (Girault) (= *Syntomosyphrum obscuriceps*) (Hym.: Eulophidae)
Nomuraea rileyi (Fungus)
Oecophylla smaragdina (Fabricius) (Hym.: Formicidae)
Ooencyrtus erionotae Ferrière (Hym.: Encyrtidae)
Ooencyrtus malayensis Ferrière (Hym.: Encyrtidae)
Ooencyrtus papilionis Ashmead (Hym.: Encyrtidae)
Ooencyrtus phongi Trjapitzin, Myartseva, & Kostjukov (Hym.: Encyrtidae)
Ooencyrtus sp. (Hym.: Encyrtidae)
Ooencyrtus submetallicus (Howard) (Hym.: Encyrtidae)
Ooencyrtus (*Tetrastichus*) *sokolowskii* Kurdj (Hym.: Eulophidae)
Ooencyrtus trinidadensis Crawford (Hym.: Encyrtidae)
Opius importatus Fischer (Hym.: Braconidae)
Opius phaseoli Fischer (Hym.: Braconidae)
Orgilus lepidus Muesebeck (Hym.: Braconidae)
Orius insidiosus (Say) (Hem.: Anthocoridae)
Orius mexidentex Ghauri (Hem.: Anthocoridae)
Orius persequens (Buchanan-White) (Hem.: Anthocoridae)
Orius sp. (Hem.: Anthocoridae)
Orius strigicollis (Poppius) (Hem.: Anthocoridae)
Orius tristicolor (White) (Hem.: Anthocoridae)
Oxyopes sp. (Arachnida: Oxyopidae)
Pachyophthalmus sp. (Dip.: Tachinidae)
Paecilomyces farinosus (Fungus)
Paecilomyces fumosoroseus (Fungus)
Palexorista inconspicuoides (Bar.) (Dip.: Tachinidae)
Palexorista solennis Walker (Dip.: Tachinidae)
Pandora gammae (Fungus)
Pandora neoaphidis (Fungus)
Paniscus ocellaris Thomson (Hym.: Ichneumonidae)
Paraglypta tubigera Kieffer (Hym.: Ichneumonidae)
Paragus serratus (Fabricius) (Dip.: Syrphidae)
Paragus tibialis Fallen (Dip.: Syrphidae)
Paraphylax fasciatipennis Ashm. (Hym.: Ichneumonidae)
Paraphylax spp. (Hym.: Ichneumonidae)
Parasierola emigrata (Rohwer) (Hym.: Bethylidae)
Paratriphleps laevisculus Champion (Hem.: Anthocoridae)
Parena nigrolineata (Chd.) (Col.: Carabidae)
Pediobius foveolatus (Crawford) (Hym.: Eulophidae)
Pediobius sp. (Hym.: Eulophidae)
Pentilia insidiosa Mulsant (Col.: Coccinellidae)
Perilissus cingulator (Morley) (Hym.: Ichneumonidae)
Phaenocarpa (Asobara) bactrocerae Gahan (Hym.: Braconidae)
Phanerotoma bennetti Muesebeck (Hym.: Braconidae)
Phanerotoma hendecasisella Cameron (Hym.: Braconidae)
Phanerotoma hindecasisella Cameron (Hym.: Braconidae)
Phanerotoma leucobasis Kriechbaumer (Hym.: Braconidae)
Phanerotoma sp. nr. *hendecasisella* Cameron (Hym.: Braconidae)
Phanerotoma planifrons (Nees) (Hym.: Braconidae)
Phanerotoma philippinensis (Ashmead) (Hym.: Braconidae)
Pheidole megacephala (Fabricius) (Hym.: Formicidae)

Philonthus turbidus Erichson (Col.: Staphylinidae)
Photoptera erythronota (Viereck) (Hym.: Ichneumonidae)
Phymastichus coffea La Salle (Hym.: Eulophidae)
Phytoseiulus macropilus (Banks) (Acari: Phytoseiidae)
Phytoseiulus persimilis Athias-Henriot (Acari: Phytoseiidae)
Plaesius javanus Erichson (Col.: Histeridae)
Plaesius laevigatus Marseul (Col.: Histeridae)
Plexippus sp. (Arachnida: Salticidae)
Plutarchia indefensa (Walker) (Hym.: Eurytomidae)
Plutarchia sp. (Hym.: Eurytomidae)
Pnigalio katonis (Ishii) (Hym.: Eulophidae)
Pnigalio pectinicornis L. (Hym.: Eulophidae)
Podisus maculiventris (Say) (Hem.: Pentatomidae)
Podothrips lucasseni (Kruger) (Thy.: Phlaeothripidae)
Polistes herebreus F. (Hym.: Vespidae)
Polycystus propinquus Waterston (Hym.: Pteromalidae)
Polycystus sp. (Hym.: Pteromalidae)
Polynema sp. (Hym.: Mymaridae)
Polyodaspis sp. (Dip.: Chloropidae)
Praestochrysis shanghaiensis Smith (Hym.: Chrysididae)
Praon abjectum (Haliday) (Hym.: Braconidae)
Praon absinthii Bagnall (Hym.: Braconidae)
Praon myzophagum Mackauer (Hym.: Braconidae)
Priochirus unicolor (Laporte) (Col.: Staphylinidae)
Pristaulacus nigripes var. *duporti* Kieffer (Hym.: Evaniidae)
Pristodoryctes striativentris Kieffer (Hym: Braconidae)
Pristomerus sp. (Hym.: Ichmenmonidae)
Pristomerus testaceus Morley (Hym.: Ichneumonidae)
Prochiloneurus io Girault (Hym.: Encyrtidae)
Prochiloneurus rex (Girault) (Hym.: Encyrtidae)
Promiscolus sesquistriatus Kieffer (Hym.: Braconidae)
Prorops nasuta Waterston (Hym.: Bethylidae)
Pseudaphycus dysmicocci Bennett (Hym.: Encyrtidae)
Pseudaphycus sp. (Hym.: Encyrtidae)
Pseudendaphis maculans Barnes (Dip.: Cecidomyiidae)
Pseudleptomastix mexicana Noyes and Schauff (Hym.: Encyrtidae)
Pseudoazya trinitalis (Marshall) (Col.: Coccinellidae)
Pseudocatoloccus nitescens Walker (Hymenoptera: Eurytomidae)
Pseudoperichaeta sp. (Dip.: Tachinidae)
Pseudoscymnus anomalus Chapin (Col.: Coccinellidae)
Psyllaephagus sp. (Hym.: Encyrtidae)
Psyttalia incisi (Silvestri) (Hym.: Braconidae)
Psyttalia (*Opius*) *fletcheri* (Silvestri) (Hym.: Braconidae)
Psyttalia sp. nr. *fletcheri* (Silvestri) (Hym.: Braconidae)
Pteromalus sp. (Hym.: Pteromalidae)
Pullus spp. (Col.: Coccinellidae)
Quadrastichus sp. (Hym.: Eulophidae)
Rhaconotus sp. (Hym.: Braconidae)
Rhinacoa forticornis Reuter (Hemiptera: Miridae)
Rhoptromeris sp. (Hym.: Eucoilidae)
Rhyzobius lophanthae Blaisdell (Col.: Coccinellidae)

Rhyzobius satelles Blackburn (Col.: Coccinellidae)
Rhyzobius ventralis Erichson (Col.: Coccinellidae)
Rogas aligarhensis Quadri (Hym.: Braconidae)
Rogas percurrens Lyle (Hym.: Braconidae)
Rogas testaceus Spinola (Hym.: Braconidae)
Scelio spp. (Hym.: Scelionidae)
Scenocharops sp. (Hym.: Ichneumonidae)
Sclerodermus domesticus Klug (Hym.: Bethylidae)
Scolothrips indicus Priesn. (Thy.: Thripidae)
Scutellista caerulea (Fonscolombe) (Hym.: Pteromalidae)
Scymnus apiciflavus Motschulsky (Col.: Coccinellidae)
Scymnus coccivora Ramakrishna Ayyar (Col.: Coccinellidae)
Scymnus roepkei Fluiter (Col.: Coccinellidae)
Scymnus sp. (Col.: Coccinellidae)
Serratia marcescens Bizio (Enterobacteriaceae)
Solenopsis geminata (Fabricius) (Hym.: Formicidae)
Spalangia endius Walker (Hym.: Pteromalidae)
Spalangia hirta Haliday (Hym.: Pteromalidae)
Spaniopterus crucifer Gahan (Hym.: Encyrtidae)
Sphegigaster brunneicornis (Ferrière) (Hym.: Pteromalidae)
Sphegigaster rugosa (Waterston) (Hym.: Pteromalidae)
Sphegigaster stella (Girault) (Hym.: Pteromalidae)
Sphegigaster voltairei (Girault) (Hym.: Pteromalidae)
Steinernema carpocapsae (Nematoda: Steinernematidae)
Stethoconus praefectus Distant (Hem.: Miridae)
Stethorus gilviforn Mulsant (Col.: Coccinellidae)
Stethynium triclavatum Enock (Hym.: Mymaridae)
Sticholotis quatrosignata Weise (Col.: Coccinellidae)
Sticholotis punctata (Crotch) (Col.: Coccinellidae)
Sturmia parachrysops Bezzi (Dip.: Tachinidae)
Sycanus collaris (F.) (Hem.: Reduviidae)
Sycanus leucomesus Walker (Hem.: Reduviidae)
Sycanus sp. (Hem.: Reduviidae)
Sympiesis sp. (Hym.: Eulophidae)
Sympiesis striatipes (Ashmead) (Hym.: Eulophidae)
Sympiesis viridula Thompson (Hym.: Eulophidae)
Syrphus spp. (Dip.: Syrphidae)
Tamarixia radiata (Waterston) (Hym.: Eulophidae)
Telenomus comperei Crawford (Hym.: Scelionidae)
Telenomus cyrus Nixon (Hym.: Scelionidae)
Telenomus pacificus (Gahan) (Hym.: Scelionidae)
Telenomus proditor Nixon (Hym.: Scelionidae)
Telenomus remus Nixon (Hym.: Scelionidae)
Telenomus sp. (Hym.: Scelionidae)
Telenomus spodopterae Dodd (Hym.: Scelionidae)
Telenomus triptus Nixon (Hym.: Scelionidae)
Telsimia nitida Chapin (Col.: Coccinellidae)
Testudobracon sp. (Hym.: Braconidae)
Tetracnemus indicus Mani & Kaul (Hym.: Encyrtidae)
Tetramorium bicarinatum Nylander (Hym.: Formicidae)
Tetramorium guineense (F.) (Hym.: Formicidae)

Tetrastichus dacicida (Silvestri) (Hym.: Eulophidae)
Tetrastichus giffardianus Silvestri (Hym.: Eulophidae)
Tetrastichus ophiusae Crawford (Hym.: Eulophidae)
Tetrastichus sokolowskii Kurdj. (Hym.: Eulophidae)
Tetrastichus sp. (Hym.: Eulophidae)
Tetrastichus spp. (Hym.: Eulophidae)
Thripobius javae (Girault) (= Thripobius semiluteus) (Hym.: Eulophidae)
Tiphia segregata Crawford (Hym.: Tiphiidae)
Toxares macrosiphophagum Shuja Uddin (Hym.: Aphidiidae)
Toxares zakai Shuja Uddin (Hym.: Aphidiidae)
Trathala flavoorbitalis (Cameron) (Hym.: Ichneumonidae)
Trathala striata Cameron (Hym.: Ichneumonidae)
Trichogramma achaeae Nagaraja & Nagarkatti (Hym.: Trichogrammatidae)
Trichogramma agriae Nagaraja (Hym.: Trichogrammatidae)
Trichogramma australicum Girault (Hym.: Trichogrammatidae)
Trichogramma chilonis Ishii (Hym.: Trichogrammatidae)
Trichogramma confusum Viggiani (Hym.: Trichogrammatidae)
Trichogramma evanescens Westwood (Hym.: Trichogrammatidae)
Trichogramma japonicum Ashmead (Hym.: Trichogrammatidae)
Trichogramma minutum Riley (Hym.: Trichogrammatidae)
Trichogramma sp. (Hym.: Trichogrammatidae)
Trichogrammatoidea bactrae fumata Nagaraja (Hym.: Trichogrammatidae)
Trichogrammatoidea sp. nr. *nana* (Zehntner) (Hym.: Trichogrammatidae)
Tricholygia bombiscum Bech (Dip.: Tachinidae)
Trichopoda pennipes (F.) (Dip.: Tachinidae)
Trichopoda pilipes (Fabricius) (Dip.: Tachinidae)
Trichospilus diatraeae Cherian and Margabandhu (Hym.: Eulophidae)
Trichospilus pupivora Ferrière (Hym.: Eulophidae)
Trioxys acalephae (Marshall) (Hym.: Aphidiidae)
Trioxys basicurvus Shuja Uddin (Hym.: Aphidiidae)
Trioxys communis Gahan (Hym.: Aphidiidae)
Trioxys equates Samanta, Tamili and Raychaudhuri (Hym.: Aphidiidae)
Trioxys indicus Subba Rao and Sharma (Hym.: Aphidiidae)
Trioxys rietscheli Mackauer (Hym.: Aphidiidae)
Trioxys rubicola Shuja Uddin (Hym.: Aphidiidae)
Trioxys sinensis Mackauer (Hym.: Aphidiidae)
Trissolcus basalis (Wollaston) (Hym.: Scelionidae)
Trissolcus mitsukurii (Ashmead) (Hym.: Scelionidae)
Trissolcus nakagawai Watanabe (Hym.: Scelionidae)
Trissolcus rudus Le (Hym.: Scelionidae)
Typhlodromus (= *Amblyseius*) *ovalis* Evans (Acari: Phytoseiidae)
Typhlodromus pyri Scheuten (Acari: Phytoseiidae)
Typhlodromips sessor DeLeon (Acari: Phytoseiidae)
Verticillium lecanii (Fungus)
Winthemia caledoniae Mesnil (Dip.: Tachinidae)
Winthemia neowinthemoides (Townsend) (Dip.: Tachinidae)
Xanthogramma scutellare Fabricius (Dip.: Syrphidae)
Xanthopimpla punctata (Fabricius) (Hym.: Ichneumonidae)
Xanthopimpla regina Morley (Hym.: Ichneumonidae)
Xanthopimpla sp. (Hym.: Ichneumonidae)
Xenoencyrtus niger Riek (Hym.: Encyrtidae)

Zagloba aeneipennis (Sicard) (Col.: Coccinellidae)
Zamesochorus orientalis Viereck (Hym.: Ichneumonidae)
Zaomma microphagum (Mayr) (Hym.: Encyrtidae)
Zaommomentedon brevipetiolatus Kamijo (Hym.: Eulophidae)
Zelus renardii Kolenati (Hem.: Reduviidae)
Zenillia heterusiae Coquillett (Dip.: Tachinidae)
Zoophthora radicans (Brefeld) Batko (Fungus)
Zygobothria atropivora Robineau-Desvoidy (Dip.: Tachinidae)
Zygobothria ciliata (Wulp) (Dip.: Tachinidae)

Index

Page numbers in **bold** refer to illustrations and tables